Textiles and Human Thermophysiological Comfort in the Indoor Environment

Textiles and Human Thermophysiological Comfort in the Indoor Environment

Radostina A. Angelova

CRC Press
Taylor & Francis Group
Boca Raton London New York

CRC Press is an imprint of the
Taylor & Francis Group, an **informa** business

CRC Press
Taylor & Francis Group
6000 Broken Sound Parkway NW, Suite 300
Boca Raton, FL 33487-2742

First issued in paperback 2017

© 2016 by Taylor & Francis Group, LLC
CRC Press is an imprint of Taylor & Francis Group, an Informa business

No claim to original U.S. Government works

ISBN-13: 978-1-4987-1539-3 (hbk)
ISBN-13: 978-1-138-89362-7 (pbk)

Visit the Taylor & Francis Web site at
http://www.taylorandfrancis.com

and the CRC Press Web site at
http://www.crcpress.com

To Peter and Alexander

Contents

Section I   Interaction of Textiles and Clothing with the Environment and Man

Section III   Mathematical Modeling and Numerical Study of the Properties of Woven Structures with Respect to Thermophysiological Comfort

List of Figures

List of Tables

Preface

Human thermophysiological comfort is one of the comfort aspects of the indoor environment. It is associated with the maintenance of thermal balance between the production of heat by the body and the heat losses dissipated to the environment. The importance of textiles for thermophysiological comfort is determined by their specific function: all the processes of heat and mass transfer between the human body and the environment are accomplished through one or more layers of textiles, which in most cases are of different material, structure, and properties. In this sense, textiles, being part of both the clothing systems and interior textiles, are the only barrier between the human body and the environment—be it indoors or outdoors.

Despite its subjective nature, to some extent, assessment of thermophysiological comfort as part of general human comfort depends largely on the design and production of textiles and clothing from an engineering point of view: from the selection of materials and their blending, through manufacturing of textiles with different properties and their finishing, to the production of garments with different applications and upholstery textiles or interior textiles like curtains and carpets.

In essence, the problems of indoor thermophysiological comfort are problems of the modern world. Many studies in the past two decades have been devoted to this issue; we have thus reached a critical mass of knowledge and experience. This observation applies particularly to the developed countries, where people spend over 90% of their time indoors. Comfort problems today are related not just to the comfort and health of people but also to their performance, which already has shown economic and social indicators: absenteeism, ineffective execution of tasks, and development of diseases associated with the indoor environment.

At the same time, the problem of comfort in the indoor environment, including thermophysiological comfort, is undoubtedly an interdisciplinary one. Assessment of the indoor environment is a complex task, which is based on the subjective feelings and preferences of occupants. Despite the existence of standards oriented to the average individual and used in the design and operation of buildings, the increasing requirements of comfort need a paradigm shift. This change calls for meeting the needs of even the most susceptible occupants of the indoor environment, while the range of activities indoors constantly expands.

It must be underlined that within the diverse activities indoors, clothing and activity are the only factors that depend on the individual and can—to some extent—be adjusted, so that the person achieves the desired thermophysiological comfort. In fact, in static conditions in terms of activity, clothing (in some cases the interior textiles as well) remains the only factor through which an individual can influence his or her thermophysiological comfort. However, a person cannot affect his thermophysiological comfort in several specific cases, i.e., when it comes to use of working or protective clothing and uniforms, or when a special dress code is required. In such cases, textiles and clothing have to ensure thermophysiological comfort indoors in the same way as is done for garments used in the outdoor environment.

This book presents a systematic study of textile structures for various applications in the indoor environment with respect to the thermophysiological comfort of the inhabitants and consists of four sections. In Section I, an overview of the role of indoor textiles and clothing as a barrier between the environment and the human body is presented; three basic aspects are discussed: *textiles and the environment, textiles as an insulation barrier,* and *textiles and the human body.* Section II presents results from a systematic experimental study on heat and mass transfer processes through woven textiles with different applications in the indoor environment. In Section III, a numerical study of the transport of air and heat through woven fabrics, by means of computational fluid dynamics, is described. An original approach for simulating the woven macrostructure as a jet system is developed and applied as well as verified with the results from the experimental study. Section IV discusses a study using the Gagge's thermophysiological model; the study presents the combined influence of indoor environmental parameters, clothing insulation, and the activity performed on the thermophysiological comfort of the occupants.

This book could not have become a reality without help. I thank God for everything that has led to the creation of this book. I thank my colleagues and students from the Department of Textiles, Technical University of Sofia, Bulgaria, for their help, suggestions, and collegiality. I extend my gratitude to my colleagues from the Centre for Research and Design in Human Comfort, Energy and Environment, Technical University of Sofia, Bulgaria, and especially to its head, Professor Peter Stankov, for creating an environment that encourages productive work. I thank the many people, institutions, and universities that gave me possibilities to generate ideas, share knowledge, and use their facilities and know-how.

My thanks are extended to CRC Press, Taylor & Francis Group, especially to Dr. Gagandeep Singh, without whose insistence and encouragement this book would not have become a reality, and to Amber Donley, for the professional project coordination and kind support. Last but not least, I thank my family, who has always been my haven and support.

Radostina A. Angelova
Technical University of Sofia

Author

Radostina A. Angelova is an associate professor in the Department of Textiles, Technical University of Sofia (TU-Sofia), Bulgaria. She is also an associate researcher at the Centre for Research and Design in Human Comfort, Energy and Environment at TU-Sofia. She is a lecturer in spinning and weaving technologies and machines, design of yarns and fabrics, smart and intelligent textiles, and protective clothing. Dr. Angelova holds an MSc degree in textiles and clothing (1994), a PhD in the technology of textile materials (2001), and a DSc in the technology of textile materials (2015). She has authored more than 100 papers and 5 books.

Section I

Interaction of Textiles and Clothing with the Environment and Man

1

Human Comfort: Thermophysiological Comfort

1.1 Role of Textiles and Clothing

The main task of textiles and clothing is to protect the human body from weather conditions with sufficient comfort in a physical and psychological sense, regardless of the type of activity. In the majority of its applications, the textile layer (or a system of layers) is a barrier against the free exchange of heat and fluids between the human body and the environment.

The development of textiles and clothing is closely linked to the development of human society and the changes in the *environment–textiles–human body* relationship. Historically, three basic stages of this interconnection can be observed (Angelova 2004):

- *First stage*: This stage refers to the period when textiles acted as a barrier between the human body and the environment. Protection from cold and solar radiation is the oldest function of textiles and clothing.
- *Second stage*: This stage refers to the first half of the twentieth century, which saw intensive research on textiles as a protective barrier against heat flows (open fire and high temperatures).
- *Third stage*: This stage refers to the period after the 1950s, which has seen an increase in the requirements of inhabitants for the comfort of the indoor environment and the higher living standards of developed societies. This augmentation of the time spent in a closed environment (including vehicles for transport), by up to 90% (U.S. Environmental Protection Agency 1989), requires deep analyses of all components of the indoor environment and their impact on human comfort, health, and productivity.

The functioning of the human body is related to temperature, both the temperature of the body and that of the environment (being it indoors or outdoors). Therefore, the body has developed in the process of evolution an

3

extremely sensitive and subtle mechanism—the thermoregulatory system—to maintain the temperature of the main organs and systems over a very narrow temperature range. However, unlike many living creatures on earth, and although people inhabit regions with extreme weather conditions, the human body cannot not survive without shelter and clothing. This is valid even more for the modern world, where people are working in extreme-temperature conditions both outdoors and indoors. Such activities cannot be carried out effectively or at all without the garments that provide thermal protection for the individual.

It has been proven that the thermal sensation decreases with age, that is, the response of the thermoregulatory system decreases; therefore, elderly people require greater protection from hypothermia.

There is a need for a comprehensive and interdisciplinary approach when describing the role of textiles and clothing in the *environment–textiles–human body* relationship. Therefore, in Section I of this book, the following aspects are presented:

- Human comfort, particularly thermophysiological comfort (Chapter 1)
- Interaction between textiles and clothing, from one side, and the environment, from the other, with respect to thermophysiological comfort (Chapter 2)
- Thermo-insulation abilities of textiles and clothing (Chapter 3)
- Interaction between textiles and clothing, from one side, and the human body, from the other, with respect to thermophysiological comfort (Chapter 4)

1.2 Human Comfort

In essence, comfort is a relative and subjective category. As defined by Hatch (1993), "comfort is freedom from pain, freedom from discomfort. It is a neutral state" (p. 26). Slater (1985) phrased comfort as "a pleasant state of physiological, psychological and physical harmony between the human being and the environment" (p. 4).

The physical aspects of comfort are related to human perception and subjective feelings of discomfort and/or pain. The factors that determine physical comfort are associated with the senses: touch, sight, hearing, taste, and smell. Touch plays a key role in people's use of textiles and clothing, together with vision (quite often) and smell (to a lesser extent and in specific cases).

From a psychological point of view, comfort reflects the individual requirements of people for clothing design, fabrics design, and colors to ensure convenience and confidence in various activities. In this sense, textiles are

TABLE 1.1

Factors Related to Textiles and Clothing That Determine
Human Comfort

	Human Comfort		
	Physical	Physiological	Psychological
Depends on	Touch	Thermal perception	Texture
	Sight	Sense perception	Color
	Smell	Movement	Design

among the factors that determine the styling, a person's relationship with others (i.e., liking), and the need to express individuality or to ensure anonymity and privacy. Very important is the role of textiles in the indoor environment, which is a combination between the individual requirements of the inhabitant and the overall architecture and interior design.

The physiological comfort of an individual is associated with maintaining the heat balance between the production of heat by the body and the heat losses. Subjective factors that determine physiological comfort are thermal perceptions of individual sensory perceptions and physical activity. The factors related to textiles and clothing that determine human comfort are summarized in Table 1.1.

Comfort, however, must be seen not only as a function of the physical properties of the material characteristics of the textile/clothing and the environment but in the complete context of human physiological and psychological response. At the same time, an individual's assessment of comfort is subjective, that is, able to "modify" and filter the objective parameters.

1.3 Physiological and Thermophysiological Comfort

1.3.1 Physiological Comfort

Physical comfort (Table 1.1) is largely a subjective factor, although influenced by receptors that are universal to the human body. Individuals have varying levels of sensitivity. Providing physical comfort is a problem of the engineering and artistic aspect of the production of textiles and clothing. *Psychological comfort* that is not less subjective depends mainly on the design, fashion trends, and other factors, mostly related to art. *Physiological comfort*, although again dependent on individual reactions and perceptions, is related strongly to the design and production of textiles and clothing in the engineering aspect: from the selection of materials and the yarn production, fabric manufacturing through weaving, knitting, and so on and their finishing, to the production of garments for different purposes, furniture production, or application of the textiles in the interior.

- Physiological comfort, determined by human sense perception, is called neuro*physiological* comfort and is associated with the following characteristics of textiles: presence of stimuli such as poke, itching, and inflammation; roughness; thermal irritation (unpleasant feeling for cold/warm); prone to static electricity.
- Physiological comfort, determined by human *activity*, depends on elasticity, weight, pressure on the body surface, and other characteristics of textiles and clothing.
- Thermophysiological comfort is a component of physiological comfort related to the thermal perception of the individual.

1.3.2 Thermophysiological Comfort

Thermophysiological comfort is directly related to the role of textiles as a barrier between the human body and the environment. It may also be defined as the ability to maintain the thermal balance between the heat generated by the body and the heat losses. The factors that determine thermophysiological comfort can be summarized as follows:

- *Factors associated with the human body*: Cardiovascular system, musculoskeletal system, central nervous system, respiratory system, digestive system, and the system for thermoregulation
- *Factors related to textiles as a barrier between the body and the environment*: Insulation ability and heat resistance, air permeability and water vapor permeability, sorption capacity/water resistance, water repellency, and speed of drying
- *Factors related to the environment*: Temperature, relative humidity, presence/absence of wind (airflow movement), presence/absence of radiation heat, and so on

1.3.2.1 Phases of Thermophysiological Comfort

Thermophysiological comfort has two different phases: in stationary and unstable (dynamic) conditions. During normal activity, the body continuously generates heat flows—dry and latent—that have to be *spread* evenly to ensure the body's thermophysiological comfort. Human skin moistens constantly, but this process is invisible in a state of thermophysiological comfort. Thus, in stationary conditions, the textile layer (or a system of layers) in contact with the body becomes part of the body's thermal regulation system.

In unstable conditions, caused by metabolic reactions or environmental conditions, the thermoregulatory system of the body causes non-uniform discharge of a moderate or significant amount of sweat by latent heat transfer. In this case, sweat and water vapor on the surface of the skin should be quickly removed by the textile layer so that thermophysiological comfort is achieved.

In most cases, the life and work of people in the indoor environment is associated with stationary activities. Intensive sport indoors or activities represent the exceptions, when the person is constantly in effect moving between outdoor and indoor environments. Therefore, the problem of thermophysiological comfort in the indoor environment is most often associated with the first phase, in which textiles are part of the whole thermoregulatory system of the human body.

1.3.2.2 Terms Associated with Thermophysiological Comfort

1.3.2.2.1 Thermal Comfort

Thermal comfort is defined in ISO 7730 (1995) as "the mental condition that expresses *satisfaction* with the surrounding environment" (Clause 7). This brief and accurate definition is quite difficult to be described by physical quantities, however.

In his classic monograph, Fanger (1972) proposed an analytical equation of thermal comfort in which different variables were involved, such as clothing, metabolism (depending on the task being performed), and environmental parameters of temperature, humidity, room airflow velocity, and so on. Fanger (1972) defined four essential conditions for an individual to be in a state of thermal comfort:

- The body is in thermal equilibrium.
- Sweat is within comfortable limits.
- The average skin temperature is within comfortable limits.
- There is no local discomfort.

Thermal comfort depends on personal factors and parameters of the indoor environment and is determined by core body temperature T_{cr} and skin temperature T_{sk} (Fanger 1967; Gagge et al. 1967).

Decades later, Parsons (2003) proved in his monograph that the comfortable limits for thermal comfort do not depend on the geographic region (cold/warm climate), age, or gender of the occupants. Of these factors, however, the sensation of cold/warm depends strongly on age, sex, weight (i.e., ratio muscle/fat), and the acclimatization of the individual.

Benzinger (1969) defined thermal comfort as a state in which there is no impulse in humans to correct the environmental conditions through some kind of behavioral reaction.

Up-to-date methods for experimental evaluation of thermal comfort in rooms (mostly ventilated and air-conditioned) include precise measurements of room airflow parameters at certain points in space and an assessment by questionnaires of clothing, activities, and personal satisfaction of the inhabitants. Modern practice involves the use of skin simulators or thermal mannequins with different features that replace human reactions and allow long and precise measurements; it would be hardly possible to make such measurements with real subjects or without ethical issues. A third possibility for the evaluation of thermal comfort is through numerical simulation of room aerodynamics with integration of thermal comfort indices, defined in ISO 7730 (1995).

Very often in the literature *thermal comfort* and *thermophysiological comfort* are used as synonyms, as the analysis of the occupants' thermal comfort in the indoor environment inevitably is based on their clothing, feelings, and reactions. Following the definition of ISO 7730 (1995), it can be clarified, however, that thermal comfort represents rather an assessment of the quality of the indoor environment, while thermophysiological comfort is directly related to the reactions of the human body, including the impact of textiles/clothing.

1.3.2.2.2 Thermal Sensation

Thermal sensation is related to the personal sensation of the individual about the temperature. It is premised on the sensitivity of human thermo-receptors and the neurological pathways used by the signals for *warmth* and *cold* to reach the brain. The actual thermal sensation is formed in the somatosensory cortex—the part of the brain believed to be responsible for the sense of touch.

Although the signals for warm and cold are related to the temperature of the skin, T_{sk}, local cooling of the upper or lower limbs can cause an overall cold feeling in the individual, which is not associated with a decrease in T_{sk} (Hensel 1981).

When a rapid change of the ambient temperature occurs, the thermal sensation is changed before the reaction of the skin and core body temperature. Gagge et al. (1967) concluded that thermal sensation is determined by body heat losses. It is therefore a logical conclusion that in the case of fluctuating environmental parameters, thermophysiological comfort should be assessed in accordance with the environmental conditions rather than with a change in core body or skin surface temperature. In the stationary conditions of the indoor environment, however, skin temperature is an important indicator for human thermophysiological comfort.

1.3.2.2.3 Thermal Discomfort

The absence of thermal comfort or thermal discomfort is an incentive for behavioral reaction. In many situations, however, the individual can change neither the indoor environment and its parameters nor the clothing. This results in decreased activity, decreased concentration, and physical discomfort (i.e., headaches). Hancock et al. (2007) showed that physiological stress causes less harm to human activity than thermal discomfort.

1.3.2.2.4 *Thermal Stress*

Thermal stress is a body condition that occurs when the environment has a very high or a very low temperature. The term is used to evaluate the effect of hot and humid climate on soldiers and the impact of the environment on active athletes. In most cases, heat stress is observed in the outdoor environment, but there are industries that are examples of work in a hot indoor environment (e.g., foundries, textile finishing departments, laundry rooms) or a cold indoor environment (e.g., freezers, warehouses) (Angelova 2007a).

1.4 Summary

The role of textile and clothing in the protection of the human body from the environment and thermophysiological comfort were presented and analyzed.

The main terms related to thermophysiological comfort were systematized and analyzed.

2

Textiles and Clothing
in the Indoor Environment

2.1 Interaction between the Human Body and the Indoor Environment

The interaction between the human body and the environment, be it outdoor or indoor, depends on a complex set of factors, which can be summarized in two groups:

- *Indoor environmental factors*: Air temperature, mean radiant temperature, room air velocity, relative humidity
- *Personal factors*: Clothing, activity, recent thermal history

To these factors, secondary factors that influence the individual's perception of heat or cold must be added (Melikov 2006). They can also be classified as environmental factors (non-homogeneity of the environmental parameters, parameters of the outdoor environment, etc.) and personal factors (adaptability, age, etc.)

It should be stressed that in an indoor environment where the heating, ventilation, and air-conditioning systems (HVAC systems) provide the prescribed standard environmental parameters, clothing and activity are the only factors that depend on the individual and can, to some extent, be managed to achieve the desired thermophysiological comfort (Angelova 2003).

In most of the cases, a person performs steady activity in an indoor environment, an activity that cannot or rarely can be changed (especially at the workplace), so the actual opportunity for the person to increase or decrease the production of heat by altering his or her activity is limited. At the same time, many jobs require the use of a dress code (offices, service centers), uniforms (restaurants, hotels), or protective clothing (hospitals, manufacturing plants), which further hinders the individual in the regulation of his or her thermal comfort by clothing change.

However, *the clothing remains the only factor that can help the individual to influence his or her thermal comfort in a particular indoor environment*. This possibility is defined as a behavioral response, which is also addressed by the thermo-physiological models that deal with the description of the thermoregulatory system of the human body (Angelova 2013).

The interaction between the environment and the human body is a complex and precise process. Usually not only one but two or more textile layers act as a barrier between the body and the environment—indoor or outdoor. Meanwhile, the thermoregulatory system is influenced by the task being performed. This complex process is presented in Figure 2.1.

Upon analysis of the process of heat transfer from the environment to the body, it is obvious that the interaction between the outside air and the textile layer(s) is a physical process of heat exchange on the basis of the existing temperature gradient. More complex is the interaction between the textile in its role of a protective barrier and the body. This process includes two parts:

- *Textile–skin* interaction, which again is a purely physical process
- *Skin–core body* interaction, which is a physiological process

As a result, the heat from the skin is transported through conduction and convection due to the circulating fluids inside the body. Hereinafter, the control of the heat flow is carried out by physiological reactions such as sweating, metabolism, and vasomotor action (Lotens 1988).

On the other hand, the activity performed by the body affects the core body temperature primarily through metabolism as well as by the nervous system. This causes a reaction in the body, which is due to the psychological and physiological processes. An opposite reaction appears in the textile layer by the released water vapor and sweat from the skin surface.

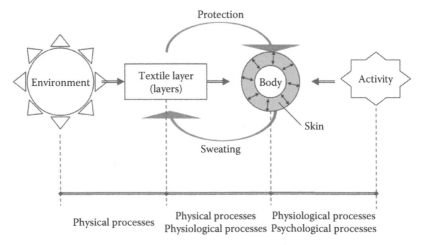

FIGURE 2.1
Effect of the environment on the human body.

2.2 Textiles in the Indoor Environment

Textiles are an excellent way to maximize the comfort and personalization of the working and living environment. They affect people's comfort indoors in two main ways (Angelova 2003):

- As a *direct insulating layer* between the body and the environment (clothing, linen, bedding, upholstery textiles)
- As an *indirect insulating layer* between the body and the environment (floor coverings, curtains, wall decorations)

Unlike clothing, which can be changed, and linen, which is easy to be replaced, other textile items in the indoor environment are relatively constant and their correct selection is essential for the quality of the living environment, including for air temperature and humidity. This fact is of particular importance when the textiles in the indoor environment are considered as one of the proven sources of hazards.

2.2.1 Risk Factors in the Indoor Environment

The phenomenon of *sick building syndrome* (SBS) was described in the early 1970s. It refers to the dissatisfaction of inhabitants with the quality of the indoor environment and a series of clinical symptoms associated with people staying in buildings. These complaints, however, had no identifiable causes. Traditional clinical studies have found that women and elderly people are more sensitive to the quality of the indoor environment, but those studies did not lead to the establishment of the mechanisms by which such sensitivity occurs and the reasons that caused it (Wyon 1994). Neither an assessment of the indoor environmental parameters from an engineering point of view nor studies on the presence of odors and chemical components could help establish the causes for SBS.

Today, SBS continues to be a subject of numerous clinical, chemical, and engineering studies; over the past decades, substantial knowledge has accumulated on the reasons for its appearance and the factors that define it. *Textiles are one of these factors.* Boestra and Leyton (1997) identified textile floor coverings and other products with surface pile as one of the risk factors for indoor air quality. Data from their study are summarized in Figure 2.2.

Gravesen et al. (1990) defined the term *macromolecular organic dust* (MOD), which included most of the known allergens. It is interesting (and disturbing) to find that allergies spread increasingly and simultaneously with an increase in SBS in Europe; they have currently reached their peak. It was assumed that the SBS was related to the clinical picture of the prevalence of allergies (Wyon 1994). Gravesen et al. (1990) showed that in the indoor environment there are too many sources of MOD, among which the authors

FIGURE 2.2
Risk factors for indoor air quality. (Data from Boestra, A.C. and Leyton, J.L., *Indoor Air*, 2, 278–283, 1997.)

defined textiles (clothing, floor coverings, curtains, and upholstery) as the most common.

According to the current understanding of air pollutants in the indoor environment, the following are the hazards:

- *Volatile organic compounds* (VOCs)
- *Microbial organic compounds* (MOCs)
- Particulate matter
- Inorganic compounds, for example, CO_2, CO, O_3
- *Semi-volatile organic compounds* (SVOCs), for example, pesticides and flame retardants

VOCs (e.g., formaldehyde, pesticides, ingredients in paints, dyes) are the most widely discussed pollutants of the indoor environment that decrease air quality (Levin 1989; Smith and Bristow 1994; Boestra and Leyton 1997).

The highest emissions of VOCs occur as a rule immediately after buildings have been constructed or a new indoor item (e.g., furniture, flooring) has been installed. The emission period of VOCs, however, could be days, weeks, or months; the duration depends mainly on the ventilation (natural or not), air temperature, and humidity (Guo and Murray 2000).

Due to their absorption ability, textiles should not be installed in the indoor environment during intensive emission of VOCs from other items.

2.2.2 Floor Coverings

Floor coverings and textiles with a pile surface are used primarily to increase the thermal and aesthetic comfort of the occupants. Smith and Bristow (1994) and Kennedy (2002) examined in detail the main advantages of floor covering textiles compared to other types of flooring (e.g., tiles, cement, wood, linoleum, bamboo). The analysis of the results and conclusions has shown that textile floor coverings

- Have the best performance in terms of thermal comfort and increase the temperature of the indoor environment in combination with *hard* flooring (e.g., parquet, cement, marble, tiles).
- Reduce noise levels during movement of occupants (steps) without decreasing the quality of desired sound (music, speech).
- Reduce fatigue and provide a safe indoor environment for young children, decreasing the risk of injury from falling.
- Have the best aesthetic qualities as they can be produced from a broad range of materials, colors, design, and touch.
- Do not reflect light.
- Can be recycled.
- Reduce living costs as they are easy to maintain.
- Decrease costs of heating.
- Eliminate the need for the use of sound-insulating materials.

When discussing the risk factors for the indoor air quality, textile floor coverings require special attention as 76% of them have a surface pile (see Figure 2.3).

The main disadvantages of textile floor coverings for indoor air quality are related to the presence of a pile surface. Even more serious is the problem with wall-to-wall carpets, which are often installed in public buildings and homes. They not only cover the whole floor, but are fixed with adhesives that contribute to increase in indoor environment pollutants. Catalli (1995) has shown that tufted carpets emit VOCs weeks after their installment in the room.

A wide range of textile materials are used in the worldwide production of textile floor coverings, but the most common are wool, polyamide, and polypropylene (Figure 2.4).

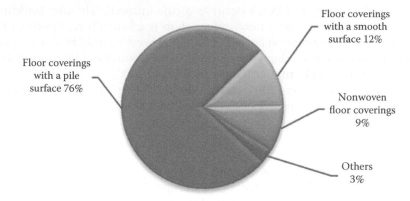

FIGURE 2.3
Distribution of world production of textile floor coverings by type.

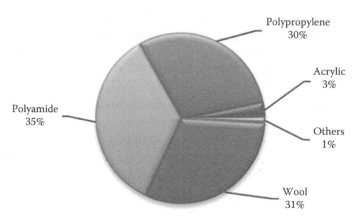

FIGURE 2.4
Types of textile materials used in the production of floor coverings.

Regardless the type of fibers used, however, textile floor coverings accumulate dust and particles, which act as allergens. A number of authors (Lewis et al. 1994; Whitmore et al. 1994; Roberts et al. 1999; Pluschke 2004) indicate carpets to be the main "reservoir" for unwanted contamination in the indoor environment. This is because they contain the greatest amount of impurities per unit area, as compared to all other surfaces in the interior, including the uncovered parts of the floor (Thatcher and Layton 1995). Therefore, a basic prescription for textile floor coverings is easy and effective cleaning. Maintenance requirements include regular vacuuming, periodic cleaning with a wet extraction, and use of cleaning products, which in turn can emit VOCs (Smith and Bristow 1994).

The type of the material used for the production of floor coverings significantly affects the amount of the accumulated dust. Tufted or needle-punched carpets with high weight and low pile from chemical fibers (polypropylene or polyamide) are particularly suited for public buildings, while wool carpets are suitable for residential buildings (Kidesø et al. 1999).

The use of wool for the manufacture of floor coverings has an advantage regarding the application of flame retardants—products that are among the main sources of pollutants (VOCs and SVOCs). Wool fibers burn slower than chemical fibers; they neither melt nor release toxic gases, so the application of flame retardants (at least for residential buildings) can be avoided (Takigami et al. 2009).

However, pesticides against moths and other insects that transform keratin introduce new volatile pollutants in the indoor air. Unacceptably high concentrations of permethrin (up to 100 mg/kg), for example, were found in residential buildings furnished with woolen carpets (Berger-Preiss et al. 2002; Butte 2003).

Wall-to-wall carpets need special attention when cleaning the next-to-wall areas, which are big sources of pollution due to difficult access (Miraftab et al. 1999). But a major risk for the health of the occupants is the ability of carpets made of synthetic fibers to accumulate mold and mildew in the case of high air humidity, whereupon the floor covering becomes a source of MOCs. Therefore, cement floors must be completely dry before carpet installation. In addition, the room has to be ventilated for at least 72 hours to reduce the emissions of VOCs, which come from both the backside of the floor covering and the adhesives used.

2.2.3 Upholstery Textiles

The comfort of upholstery textiles is normally associated with their tactile properties, especially when the uncovered parts of the human body are in contact with "soft" furniture (Pierce 1930; Howorth and Oliver 1958). In fact, tactile properties are among the factors for physical and neurophysiologic comfort of the individual. Pierce (1930) defined a number of factors that affect the tactile properties of textile fabrics, like drapeability, flexibility, thickness, densities (porosity), and elasticity.

Out of these tactile factors, however, the comfort related to furniture covered with fabrics is determined by the thermophysiological comfort of the individual, which in turn depends on the behavior of the textile barrier (the upholstery textile) between the body and the furniture (Barker 2002).

Due to the need to ensure a pleasant touch, the essential part of furniture fabrics is produced with a pile surface. Such a structure is not resistant to abrasion, which is a basic requirement for upholstery textiles, since the replacement of the furniture surface fabric is too expensive. Therefore, upholstery textiles are commonly made of chemical fibers: polyester, polyamide, polypropylene, acrylic, and so on. These fabrics are easy to maintain, stain resistant, and significantly more resistant to abrasion than fabrics made of natural fibers.

Upholstery textiles are not a risk factor for indoor air quality in terms of emissions (McCullough et al. 1994). The plush surface, however, is associated with the same problem of accumulation of dust as the floor covering textiles. Although the surface pile of upholstery textiles is significantly lower than that of carpets, for example, it also becomes a source of VOCs, MOCs, and so on as a result of the accumulation of dust and other particles from allergenic sources of harmful substances in the indoor environment.

Among the tasks of upholstery textiles is that they should provide thermophysiological comfort for the individual. Their role is similar to that of clothing, but upholstery textiles are considerably more complicated, since the textile is only a surface layer, for example, of a chair or a mattress. However, furniture fabric is expected to provide the necessary heat transfer and fluid motion between the body and the furniture, exactly as in the case of clothing.

2.2.4 Bedding and Blankets

Bed linen and blankets are associated primarily with two types of hazards in the indoor environment: MOCs and particulate matter. In specific cases, SVOCs can be also found, usually for relatively short periods of time.

The main task of bedding and blankets in the indoor environment is to provide thermophysiological comfort for the individual at rest (Mizuno et al. 2005). As the human body produces minimal heat in this state, textiles for bedding and blankets are required to have higher thermal insulation capability (Muzet et al. 1984). Primarily natural fibers are used (wool, cotton, down, feathers), followed by chemical fibers and blends.

Like fabrics for apparel, textiles for bedding and blankets must absorb and transport the water vapor emitted by the human body during rest and sleep. Therefore, cotton, linen, and silk fibers are particularly suitable, while use of their mixtures with chemical fibers or the use of 100% synthetic fibers deteriorates the thermophysiological comfort of the human body (Umbach 1986). Critical is the role of textiles for bedridden patients, adults with a high degree of immobility, and newborns, as they spend a substantial part of their time in bed (Worfolk 1997; Holland et al. 1999). To avoid discomfort and an appearance of bedsores, textiles used for bedding and blankets must ensure thermophysiological comfort, to allow transport of air, heat, and moisture as well as provide a nice touch to the uncovered parts of the body, and not be a cause of allergic reactions.

Bedding and blankets are important items for the accumulation of particulate matter in the indoor environment, which, in combination with high air humidity and improper ventilation, leads to the development of MOCs and microorganisms (i.e., mites) that cause allergic diseases.

2.2.5 Curtains and Screens

Besides enhancing aesthetic and psychological comfort, the main task of curtains in the indoor environment is to increase thermal comfort and improve the acoustic environment through the suppression of sounds from the outdoor environment. Special cases are textiles in the role of screens in public buildings (hospitals, offices) and residential premises.

Drapes and curtains are made of a whole range of natural and chemical fibers and filaments, but synthetic materials dominate (Angelova 2003). The main reason for this is their ease of maintenance, good drapeability, and resistance to UV rays. Like furniture fabrics, curtains and screens accumulate pollutants and odors; they therefore require regular maintenance. In buildings where this is not possible, the use of blinds or special windows glasses is recommended.

2.3 Summary

The interaction between the human body and the environment (with a focus on the indoor environment) was analyzed.

The controversial impact of textiles on the indoor environment as a source of comfort and hazards was detailed and analyzed.

It was shown that the choice of textiles in the indoor environment is a complex task that goes beyond aesthetic features and requires an assessment of a number of factors: the role of the textile product, the materials used for its production, the opportunities for effective cleaning, the risk for emission or re-emission of indoor air pollutants, detergents required for its maintenance, the risk of the release of additional pollutants in the form of volatile substances, and so on.

3

Thermal Insulation Properties
of Textiles and Clothing

3.1 Hierarchical Structure of Textiles

The *hierarchical structure* implies relationship and mutual dependence. Textiles obviously have such a structure, as textile materials (fibers, filaments) determine the properties of the threads (staple yarns or polyfilaments), which, in turn, determine the properties of the fabrics (of course, there are other factors, as well, i.e., the manufacturing parameters). Fabrics, as end products of the textile industry, determine to a great extent the properties of apparel or behavior of textiles in the indoor environment.

Therefore, the term *hierarchical structure of textiles* is used hereinafter with its levels:

- *Microstructure*: The fibers used for spinning the yarns or the monofilaments that form the polyfilaments
- *Mesostructure*: The staple fiber yarns or the polyfilaments
- *Macrostructure*: The fabric itself being it woven, knitted, or nonwoven, with its geometrical, mass, and structural characteristics, texture, design, finishing, and so on

In the literature, the terms *micro-, meso-, and macrostructure* are usually associated with composite materials and are not used frequently in the bibliography, devoted to traditional textiles. Their use, however, shows clearly the order and logical links in the construction of the textiles. Thus, a summary assessment of each structural level can be made, without going into the specific diversity behind it.

The hierarchical organization of the textiles makes the problem of testing and evaluation of their thermal insulation properties quite complex—in terms of evaluation of both fluid permeability and heat transfer processes.

In the particular case of movement of fluid in through-thickness direction of a woven structure, the flow passes not only through the pores of the macrostructure (the spaces between the threads), but also through the pores of mesostructure (the spaces between the fibers or monofilaments). Only in the case of monofilaments used as warp and weft threads, the fluid does not pass through the mesostructure. As for the transmission of heat and moisture, the impact of the microstructure should also be considered. According to Lomov et al. (2010), the influence of the nanostructure has to be also taken into account if a treatment with nanoparticles is performed or the woven structure is used as a phase for coating with a nano layer.

3.2 Heat Transfer through a Textile Layer

The heat transfer through a textile layer of any kind (knitted, woven, or nonwoven) differs from the transfer of heat through a solid body and includes the three mechanisms of heat transfer: *conduction, convection, and radiation*. This is due to the presence of air both in the mesostructure (between the fibers in the yarn or monofilaments in the polyfilament) and in the macrostructure (between the threads). In the nonwoven macrostructure, it is due to the presence of air between the fibers in the tissue (Angelova 2003).

The process of heat transfer through a textile layer or clothing includes the following:

- Dry heat transfer by conduction
- Dry heat transfer by convection
- Dry heat transfer by radiation
- Latent heat transfer by diffusion of water vapor
- Latent heat transfer by diffusion of sweat (liquid)

When the insulation properties of textiles and clothing are discussed in terms of thermophysiological comfort, only the first four mechanisms of heat transfer must be taken into account. The appearance of the fifth mechanism—liquid on the skin—is an indication for thermal discomfort (Angelova 2004).

The total heat transfer through a textile layer and/or clothing can be presented as follows:

$$Q = Q_d + Q_e \tag{3.1}$$

where:
Q_d is the dry heat transfer, W
Q_e is the latent heat transfer, W

3.2.1 Dry Heat Transfer

The textile structures are anisotropic, and the dry heat transfer in them differs from that in the solid bodies (Fricke and Caps 1988; Haghi 2002). At first, the yarn can be a mixture of fibers with different heat transfer coefficient. Second, air exists between the fibers in the mesostructure of woven and knitted textiles, as well as in the macrostructure of all types of textiles. Therefore, the dry heat through a textile layer is transferred via conduction, convection, and radiation.

3.2.1.1 Conduction

The conduction depends on the existence of a temperature gradient: particles of a substance with higher energy seek to reduce their energy by transferring it to particles with lower energy in colder areas. It is performed in solids, liquids, and gases, and therefore, conduction can be observed between fibers, yarns, fabrics, and the air between them, that is, in the micro-, meso- and macrolevel of the textile structure, as well as between the macrostructures involved in the construction of a garment.

The heat flow q, W/m^2, is proportional to the temperature gradient according to the differential form of the Fourier's law:

$$q = -\lambda A \frac{\partial T}{\partial x} \tag{3.2}$$

where:
$\partial T/dx$ is the temperature gradient in the heat flow direction, K/m
λ is the thermal conductivity of the material, W/mK
A is the area, m^2

3.2.1.2 Convection

The convection is associated with the directed movement of micro-volumes in a fluid medium due to the presence of a temperature gradient. It is performed in liquids and gases; therefore, convection in textile structures is due to the presence of the air between the fibers in the mesostructure, the air between the threads in the macrostructure, as well as the different macrostructures in the structure of clothing. It is expressed by the Newton's law of cooling:

$$q = h(T_s - T_f) \tag{3.3}$$

where:
T_s is the surface temperature, K
T_f is the fluid temperature, K
h is the convective heat transfer coefficient

Only the natural convection is considered in the indoor environment, unless it is not related to work activities close to fans and other devices that lead to forced convective transfer through the textile layer(s). In the outdoor environment, however, the heat exchange through textiles and clothing is due to both natural and forced convection.

The particular surface area of the textile affects the boundary layer between the fluid and the textile's mesostructure. Therefore, the texture, surface treatment, type of yarns (threads), and so on affect the convective heat transfer.

3.2.1.3 Radiation

Heat transfer by radiation is based on the distribution of quantum particles or photons. It is connected with the process of heat emission from the surface of hot bodies, which can be performed even in vacuum.

3.2.2 Impact of the Hierarchical Structure on the Dry Heat Transfer

Figure 3.1 shows the scheme of the heat transfer process from a layer with higher temperature to a layer with lower temperature.

If the layer with higher temperature is the surface of the skin, and the layer with low temperature is the air of the environment, surrounding the body, the heat exchange between the two layers is performed through conduction, convection, and radiation. If two textile layers exist between the skin and the ambient air, the processes occur identically in each layer: the surface of the textile layer emits heat by radiation, while conduction and convection appear through the air layers located either between the skin and the textile layer, or between the two textile layers.

If the type of the textile macrostructure is specified, the process of heat transfer can be detailed even further, as shown in Figure 3.2. The scheme is

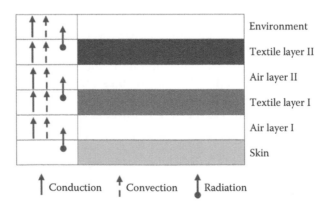

FIGURE 3.1
Dry heat transfer from the body to the environment through textile layers.

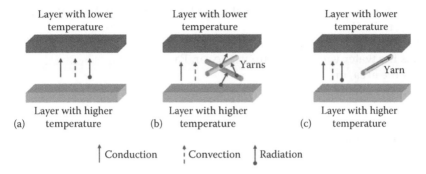

FIGURE 3.2
Dry heat transfer through a textile layer: (a) heat transfer between layers with different temperature, (b) heat transfer with a presence of a textile between the layers, and (c) heat transfer alongside a yarn.

valid for *human body–textile* heat transfer, *textile–textile* heat transfer, as well as *textile–environment* heat transfer.

If the temperature of the surrounded air is higher than that of the skin surface, these processes can be considered analogous in the opposite direction.

The heat between two layers with different temperature (Figure 3.2a) is transferred by conduction, convection, and radiation. In the case a textile is placed between the two layers, that is, a woven fabric, the process of radiant heat transfer is complicated (Figure 3.2b). The radiation heat is transferred from the layer with a higher temperature to a thread of the textile macrostructure, followed by heat transfer from thread to thread in the macrostructure, and then, from a thread of the textile layer to the layer with a lower temperature. At the same time, heat transfer occurs by means of conduction as well, if considering the length of a thread (Figure 3.2c). Taking into account the textile structure on meso-level, it is known that air exists between the staple fibers in the yarn, so convective heat flow can also appear, though in a very small amount.

3.2.3 Latent Heat Transfer

To provide thermal comfort, the textile macrostructures in the indoor environment must also ensure heat exchange through latent heat transfer. The latent heat transfer $R_{e,f}$ through a single fabric layer can be expressed by the evaporation resistance, which also includes the resistance of the boundary air layer (McCullough et al. 1989):

$$R_{e,f} = R_{e,t} - R_{e,a}, \text{ for} \qquad (3.4)$$

$$R_{e,t} = \frac{(P_s - P_a)A}{Q_e} \qquad (3.5)$$

where:
$R_{e,t}$ is the total resistance to heat transfer via latent heat of the whole clothing system and the closest air layer of the environment, m^2Pa/W

$R_{e,a}$ is the resistance to heat transfer via latent heat from the outer air layer, m²Pa/W
P_s is the water vapor pressure on the skin, Pa
P_a is the water vapor pressure in the air, Pa
A is the area, m²
Q_e is the latent heat flow, W

Equation 3.5 is similar to the equation for resistance to dry heat transfer through a single textile layer R_t:

$$R_t = \frac{(T_{sk} - T_a)A}{Q_d} \tag{3.6}$$

where:
T_{sk} is the skin temperature, °C
T_a is the air temperature, °C

The relationship between $R_{e,t}$ and R_t is given by the moisture permeability index i_m (Degen et al. 1992):

$$i_m = \frac{1}{LR}\frac{R_t}{R_{e,t}} \tag{3.7}$$

The term LR is the Lewis relation, which compares the coefficients for transfer of mass h_{mass} and heat h:

$$\frac{h_{mass}}{h/c_p} = 1 \tag{3.8}$$

where:
c_p is the specific heat capacity, J/(kg °C)

3.2.4 Impact of the Hierarchical Structure on the Latent Heat Transfer

Figure 3.3 presents the mechanism of latent heat transfer using the same treatment of Figure 3.1.

The transfer of latent heat between the textile layers is performed by diffusion, and the outer textile layer emits heat to the environment through convection. It should be particularly noted that the transfer of both dry and latent heat depends strongly on the moisture content of the textile layer, that is, mainly on the sorption capacity of the microstructure. The ability of the textile macrostructure to provide a smooth transfer of water vapor and a liquid (sweat) from the surface of the body to the environment determines thermophysiological comfort in particular indoor or outdoor environment.

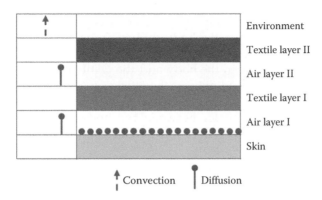

FIGURE 3.3
Latent heat transfer from the body to the environment through textile layers.

3.3 Transfer of Fluid through a Textile Layer

Figure 3.4 shows a scheme of fluid flow with velocity u_0 and flow rate Q_0 through a porous screen of a woven structure when a pressure gradient exists in the transverse direction of the specimen. Part of the flow is moving with velocity u_1 through the pores between the threads of the macrostructure and its characteristics depend on the number and size of these pores (Angelova 2012).

Another part of the flow, however, passes with velocity u_2 through the pores between the fibers in the threads (if they are staple fiber yarns or polyfilaments). It is not possible to estimate experimentally the ratio between the flow rate through the pores of the macrostructure and the pores of the mesostructure. Analysis of the process can be done, however, through numerical simulation, using the advantages of modern *computational fluid dynamics* (CFD) (see Section III).

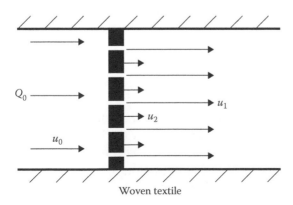

FIGURE 3.4
Transfer of fluid through a screen of a porous woven structure.

The movement of air through a textile layer is observed in a number of situations, especially when the body moves. In general, body movement increases the transport of air through the textile barrier, which in turn extracts heat from the zone between the body and the textile layers, or between the individual textile layers, and reduces the efficiency of the insulating capability of the clothing.

The increase in air transfer between the body and the environment through a textile layer(s) is due to a combination of two factors (Angelova 2008):

- The increased air flow velocity
- The effect of *pumping* of the textile layer while bending during body movement

It was found in the monograph of Goodfellow (2001) that the insulating ability of a business suit decreased by 52% if the person, wearing it, moved at a speed of 90 steps/min (approximately 3.7 km/h).

3.4 Thermal Insulation of Textiles and Clothing

As it has been already mentioned, the heat transfer from the human body through a system of textile layers to the environment is a complex process. In order to facilitate the calculations, Gagge et al. (1941) introduced a simplified term *thermal insulation of clothing* I_{cl}, which describes the thermal insulation properties of textiles.

In ISO 9920 (1995), the thermal insulation of clothing is defined as follows:

$$I_{cl} = \frac{\bar{T}_{sk} - \bar{T}_{cl}}{H}, \frac{°Cm^2}{W} \tag{3.9}$$

where:
$\bar{T}_{sk}$ is the mean skin temperature, °C
$\bar{T}_{cl}$ is the mean temperature of the outer clothing surface, °C
H is the heat losses per square meter of the human skin, W/m²

In addition to the standard unit (Equation 3.9), two other units are used for textiles thermal insulation: *clo* and *tog*. The *clo* unit was introduced by Gagge et al. (1941):

$$clo = 0.155 \frac{°C\,m^2}{W}$$
$$(3.10)$$

$$clo = 0.648 \frac{°C\,m^2 s}{cal}$$

The *tog* unit was introduced by Peirce and Rees (1946):

$$tog = 0.100 \frac{°C\,m^2}{W}$$
$$(3.11)$$

$$tog = 0.418 \frac{°C\,m^2 s}{cal}$$

The relationship between the two units is

$$1\,clo = 1.55\,tog \qquad (3.12)$$

The thermal insulation of n textile layers is a sum of the thermal insulation I_{layers} of the single layers (ISO 9920, 1995):

$$I_{cl} = \sum_{i=1}^{n} I_{layers,i} \qquad (3.13)$$

3.5 Factors That Influence the Thermal Insulation of the Textiles

Clothing for everyday use are made of fabrics (woven or knitted), and most often between the body and the environment there are two or more textile layers, including nonwovens as a hidden (internal) layer. Many studies have been devoted to the problem of *thermal insulation properties of textiles*, but readers can sometimes be confused by the published results because the publications concern two different problems: insulation properties of single textiles and insulation properties of clothing (systems of textile layers). Several research studies have reported that the thermal insulation properties of clothing depend mainly on the thickness of the textile system (Corbelini 1987; McCullough et al. 1989; Havenith 2002; Roberts et al. 2007, etc.). This is due to the fact that the role of the air layers, sandwiched between two textile layers, is as important as the characteristics of the fabrics (type of fibers, yarns, weaves, knits, etc.). The same is valid for the resistance to water vapor transmission: the volume of the air layers is much greater than the volume of the *solid part* (i.e., fibers), so that the resistance to diffusion transport of water vapor depends mainly on the air layers (air volume) between the textile layers in the clothing (McCullough et al. 1989).

The situation is different when only one textile layer is a subject of a study (Ukponmwan 1993; Schacher et al. 2000; Angelova 2012). In this case, the type of the fibers, their distribution in the yarn, the type of weave or knit used, finishing, and so on affect the macrostructure and therefore may affect its thermal performance: first, because the convective heat transfer cannot be ignored on a macro-level, and second, because the diffusion properties of a thin layer are different. The properties of the fibers used, the type of the fabric (woven, knitted, nonwoven), and the characteristics of the meso- and macrostructure are essential. All these factors can affect the heat transfer, moisture sorption, air permeability, and so on.

Fanger (1972) determined the porosity and especially the thickness of the fabrics as the main factors that affect the thermal insulation abilities of textiles. Based on textile publications from 1940s and 1950s, he concluded that the type of fibers (cotton, wool, chemical fibers, glass fibers, etc.) had no considerable influence on these properties. Possible influence of structural, geometric, and mass characteristics of the macrostructure on its thermal insulation properties was not discussed. The dominant influence of the fabric thickness on its thermal insulation properties was also commented in later publications (i.e., Ukponmwan 1993).

In terms of the complexity of the heat exchange processes between the human body and the environment through a textile layer, made of various fibers, with different structure and properties, it is logical to discuss a wider range of factors affecting these processes. And this is motivated by the practical standpoint that the rejection of the influence of other characteristics of the textiles, besides thickness and porosity, is an assumption that seems difficult to be accepted after the development of new fibers, yarn, and fabric structures and finishes after 1970s.

However, a review of the literature after 1990s showed that several authors continue to conduct research on the impact of different characteristics of textiles on their insulating properties. The overview work of Ukponmwan (1993) gave a good idea of the research in the field in the second half of the twentieth century. The development of new fiber was discussed in the work of Umbach (1993) and Schacher et al. (2000), where the transfer of moisture and the heat-insulating properties of fabrics of polyester microfibers were investigated. Structural characteristics of the fabrics and their air permeability were studied in correlation in Dubrovski and Sujica (1995). Air permeability factor, but as a function of the type of the fibers and their ability of moisture absorption and moisture preservation, was also studied by Gibson et al. (1999). The transfer of moisture in materials for garments was investigated in the works of Barnes and Holcombe (1996). Lotens and

Havenith (1991) proposed a theoretical model to calculate the thermal insulation properties of clothing.

The topic remains relevant today, and the studies of Cay et al. (2004), Daukantienė and Skarulskienė (2005), Xu and Wang (2005), Wilbik-Halgas et al. (2006), Militky et al. (2010), and Fan and He (2012) are just examples of the more and more detailed research on the relationship between the textile characteristics and the properties of textiles that determine the human thermophysiological comfort.

On the basis of the literature survey, a systematic *classification* of the factors that determine the thermal insulation properties of textiles is developed. It includes four groups of factors, as listed below:

1. *Structural factors* are related with the type and the structure of the textile:
   - Type of the macrostructure includes woven, knitted, and nonwoven.
   - Characteristics of the macrostructure are densities, thickness, mass per square meter, volume mass, porosity, and so on.
   - Characteristics of the mesostructure are type of threads (staple fiber yarns, filaments), linear density, twist, compact density, and so on.
   - Specific heat of the microstructure.
2. *Surface factors* are related to the surface of the macrostructure:
   - Texture, depending on the weave or knitted pattern.
   - Hairiness of the mesostructure, related to fiber types and structure of the threads.
   - Surface, depending on the finishing processes applied on the macrostructure.
   - Contact surface between the macrostructure and the human body.
3. *Factors related to dry and latent heat transfer*:
   - Heat transfer between the microstructure and the air in the mesostructure.
   - Heat transfer between the mesostructure and the air in the macrostructure.
   - Heat losses by conduction between the human body and the textile.
   - Heat losses by radiation from both the skin surface and the surface of the outer textile layer to the environment.
   - Heat losses by convection between the body and the nearest textile layer and between the layers in the system (clothing).

- Heat losses by diffusion from both the body surface and the surface of the outer textile layer to the environment
- Heat storage due to absorption or adsorption of moisture from the surface of the textile layer(s).

4. *Factors related to environmental parameters*:
   - Air temperature.
   - Relative humidity.
   - Air velocity.

It is clear that some of the above factors may have negligible impact on the thermal insulation properties of the textiles in terms of practical assessment of thermophysiological comfort. However, the increasing importance of thermophysiological comfort as part of the overall comfort in the indoor environment requires systematic research on a wide range of textiles and clothing to be performed in order such conclusions to be made.

3.6 Methods for Determination of Thermal Insulation Properties of Textiles

There is a variety of methods for measuring the thermal insulation properties of textiles and clothing. Detailed description of a substantial part of them can be found in the review of Ukponmwan (1993). The main measurement methods are the following:

- *Guarded hot-plate method*: The sample is placed between two metal plates with different temperatures and the heat flow through the sample is measured.
- *Cooling method*: A hot body is wrapped with the sample, whose outer surface is exposed to air impact and the degree of cooling of the body is measured.
- *Constant temperature method*: A hot body is wrapped with the sample, but it is measured for the energy needed to maintain the temperature of the hot body constant.

Among these appliances and devices, thermal mannequins are with particularly high application. The thermal mannequin allows the measurement of thermal insulation properties of both individual garments and the whole clothing ensemble (from underwear to outerwear). This device has the actual size of a person and is equipped with systems for temperature control, sweating, breathing, and so on.

Field measurements with participation of real subjects are special way for the evaluation of the thermal comfort in the indoor environment. During field measurements, not only the thermal insulation properties of textiles are assessed (usually with the help of questionnaires) but also the activity and the influence of environmental parameters.

3.7 Thermophysiological Comfort and Special Textile Materials and Garments

Today, textile materials are among the most important high-tech materials. They are increasingly being used in innovative items. Thanks to their characteristics, obtained during complex industrial processes, the range of textiles application extends from vehicle and aircraft construction, structural and civil engineering, to agriculture, environmental engineering, sports and recreation, medicine and digital systems. Even traditional textiles for household and clothing applications are provided with more and more additional functions.

High-tech solutions for improving the role of textiles and clothing as insulating barrier between the human body and the environment can be found in the *intelligent textiles*, which are applied for the first time as a concept in everyday clothing (Angelova 2006).

Although the technical report CEN/TR 16298 (2011) uses *smart* and *intelligent* textiles as synonyms, they are not identical. The essential difference is that intelligent textiles, unlike the smart textiles, do not incorporate electronic components and devices and do not process information electronically (Park and Jayaraman 2003; Van Langenhove and Hertleer 2004).

Examples of intelligent textiles are textiles with phase-change materials, color-changing textiles, microporous textiles, bioactive textiles and others. Of course, they can be equipped with electronic components, sensors, actuators, elements for transmission of information, thus obtaining the behavior of smart textiles as well.

3.7.1 Silver Fibers

The modern medicine has found silver to be a very effective, natural, and antimicrobial element, which eliminates many common types of bacteria and fungi that can provoke infections and unpleasant odors. Silver is used in wound dressings, bandages, underwear, and apparel textile materials. Its electro-conductive and electrostatic properties are suitable for application in textiles, which can create and maintain certain microenvironment around the human body, that is, in linings of outwear garments (Foulger and Gregory 2003).

Yarns with silver content are produced in different varieties:

- *Core yarns*: They consist of a core made of ultra-thin silver filament, covered by natural or man-made fibers.
- *Staple fiber yarns*: Man-made fibers or blends of man-made and natural fibers are spun and subsequently coated in pure silver.

Being one of the most conductive elements, silver offers several advantages, when incorporated in textile and clothing. Its molecules help the process of heat transfer from the skin to the surrounded air through the textile, thus providing better thermophysiological comfort for the person. The combination between the antimicrobial effect of the silver and its heat transfer properties makes the silver fibers very appropriate for application in medical products like wound dressing and bandages. Thus, the treatment time is shortened; better hygienic conditions and thermophysiological comfort for the patient are provided.

3.7.2 Carbon Fibers

Carbon fibers (alternately called graphite fibers) consist of extremely thin fibers of about 0.005–0.010 mm in diameter and composed mostly of carbon atoms. The carbon atoms are bonded together in microscopic crystals that are more or less aligned parallel to the long axis of the fiber. The crystal alignment makes the fiber incredibly strong for its size. The carbon yarns are formed by several thousands of carbon fibers, which are twisted together. The carbon yarns can be furthermore woven into a fabric.

Recently, carbon fibers were applied as heaters of clothing systems (Bryant and Colvin 1992; Zimmerli 2000). Carbon fabrics and carbon/glass composite fabrics appear to be ideally suited for relatively large heated assemblies such as vests and casualty bag liners. They can be incorporated as well as electrical heating elements into personnel clothing and protective clothing.

Carbon fibers are far more flexible than the metal conductor heating elements, used traditionally for heating of clothing systems. Carbon fiber panels do not have the unwelcome thick wires that are generally the main disadvantage of heated clothing. The main advantage of the carbon heating panels is that their incorporation in clothing does not require designed from scratch patterns and designs (Zimmerli 2000). Traditional clothing lines can be equipped with these inserts with only minor modifications. The control of the temperature of the microenvironment around the human body is performed by a heating system, which is connected to standard sized batteries or Li-Ion rechargeable batteries.

3.7.3 Phase-Change Materials

One alternative of the clothing for outdoor environment is the garments with *phase-change materials* (PCM): substances that are able to accumulate or release heat at a certain temperature (Foulger and Gregory 2003; Ilmarinen 2005).

To prevent leakage during the liquid phase of the material, PCM is enclosed in tiny, impervious microcapsule. Microcapsules can be incorporated either in the microstructure of the materials (fibers) or in the meso- and macrostructure (Angelova 2007b).

Phase change materials are used in the indoor environment as they are included in wall and floor boards to store or release heat and assure comfortable ambient temperature in the room. They have also been incorporated into cups, glasses, and tableware to keep foods and beverages at a desired eating temperature for extended periods of time.

Incorporated into textiles, PCM have the following advantages, compared to traditional textiles (Bryant and Colvin 1992; Pause 1995, 1998):

- Garments with PCM adapt to the temperature of the microenvironment around the human body.
- The overheating is avoided.
- There are less heat losses due to latent heat transfer.
- The risk of decreasing the temperature of the microenvironment below the thermo-neutral for the particular activity is avoided.
- Active temperature regulation is provided.

3.7.4 Application of Special Textile Materials

Special materials that provide thermophysiological comfort are used in textiles and clothing with different functions (Shim and McCullough 2000; Shishoo 2000; Kyeyoun et al. 2004; Salaun et al. 2010).

3.7.4.1 Active Wear

Classical active wear garments cannot always ensure the thermophysiological comfort of the body. The heat produced by the body in laborious activity is often not discharged into the environment in the required amount. The result is thermal stress situation. On the other hand, the human body produces less heat during the rest between activities. Considering the same heat release, hypothermia can occur. Application of PCMs in clothing supports thermophysiological comfort, avoiding the thermal shocks and thermal stress. The result is an increase in the work performance under high stress.

3.7.4.2 Automotive Textiles

In summer, the temperature inside the passenger compartment of an automobile can increase significantly when the car is parked outside. The use of special materials and smart textiles for the car seats and compartment can reduce considerably the energy costs for ensuring the thermophysiological comfort of the passengers.

3.7.4.3 Outdoor Sports Clothing

The main aim of the special textile materials is to protect the core body from low temperatures and hypothermia. They are used in garments for outdoor and winter sports as well as for sports performed in severe conditions (mountain climbing, cave exploration, etc.): jackets and jacket linings, boots, golf shoes, running shoes, socks, and gloves for ski and snowboard. It is expected special textile materials to be used in the production of blankets, sleeping bags, mattresses, and mattress pads.

3.7.4.4 Lifestyle Apparel

PCMs are also used in consumer and designers products: elegant fleece vests, hats, gloves, and rainwear of companies like Gucci, Pierre Cardin, Polo Ralph Lauren, Prada, Schoeller, Freudenberg, and others.

3.7.4.5 Aerospace Textiles

Smart textiles for maintenance of thermophysiological comfort are used in space suits and gloves (or their linings) to protect astronauts from higher temperature fluctuations while performing extra-vehicular activities in space.

3.7.4.6 Medical Textiles

The use of special textile materials to provide thermophysiological comfort is still not widespread since the high cost of the items (with PCMs, for example) is in conflict with the requirements for use of the disposable textiles or textiles, which are subjected to special sterilization and disinfection processes. However, such a textile materials can improve the thermophysiological comfort of surgeons, being incorporated in the fabrics for gowns, caps, and gloves. The thermophysiological comfort of patients can also be significantly improved even in the operation table when special sheets and covers maintain the body temperature and keep the patient warm enough. Mattress covers, sheets, and blankets with special textile materials can be useful for patients in thermal stress conditions.

It is expected that in the future smart and intelligent textiles can create thermal environment that is individual for each person. But such items must have the ability to continuously monitor the reactions of the body (mainly related to metabolism) and maintain the individual thermophysiological comfort with materials that can reduce fluctuations in body temperature, change their air permeability, vapor permeability, and so on.

Van Langenhove and Hertler (2004b) proposed in their study an introduction of individual comfort index for each particular case of indoor environment, activity, clothing, and so on. However, such a database is not developed or supported.

3.8 Summary

The process of transmission of fluids and heat transfer through textile layer(s) or clothing was analyzed. The effect of the impact of the hierarchical structure of textiles on dry and latent heat transfer was described in detail.

The factors, which determine the thermal insulation properties of textiles, were summarized. A classification with four groups of factors, influencing the thermal insulation properties, was developed.

An overview of publications in the field of thermal insulation properties of textiles was presented. The application of special textile materials and products, related to thermophysiological comfort, was also discussed.

4

Interaction between Textiles and Clothing and the Human Body

When used in apparel, textile macrostructures have a specific and very important function: to serve as a mobile microenvironment for the human body. In this role, textiles and clothing support the thermoregulation processes of the human body and provide thermophysiological comfort.

Along with this, the textile macrostructures are part of the indoor environment in the role of interior textiles, which also have an impact on occupants' comfort, either through direct contact with the human body (thermophysiological comfort) or influencing the parameters of the indoor environment (thermal comfort, acoustic comfort, indoor air quality).

In this chapter, the general considerations of the interaction between textiles/clothing and the human body are presented. The mechanisms of thermoregulation of the human body are detailed in Section IV.

4.1 Regulation of Body Heat

In conditions of *thermal neutrality* and *repose*, human body seeks to regulate its temperature around 37°C. This temperature increases during the day (usually around ±0.8°C) reaching its peak late at night and decreases again till the morning because of circadian rhythm. The circadian rhythm is controlled by the hypothalamus and provides 24-h adaptation of the human organism to environmental changes (Kolodyazhniy et al. 2011). Exercising can provoke an increase in body temperature, which may even exceed 40°C, that is, during a marathon (Havenith 2003).

The greater part of body's energy is used to maintain the temperature of the inner part, the *core of the body*, about 37°C. For this purpose, the body strives for balance between heat generation and heat losses. The difference between them is defined as *body heat storage* (Aoyagi et al. 1995). Two main processes occur:

- In a cold environment, the body reduces the heat loss and increases the heat production.
- In an environment with higher temperature, the body reduces the heat production and seeks to transport the surplus heat by increasing the heat losses.

If the body becomes too hot, its heat censor (the hypothalamus) initiates two mechanisms for regulating the body's temperature (Kosaka et al. 2004):

- *Vasodilatation*: It is expressed in increase in the blood flow close to the skin surface. The result is faster and easier cooling of the blood flow and consequently of the internal organs.
- *Sweating*: It is related with water vapor and sweat emission from the skin to the surroundings.

If the body becomes too cold, the cold sensors, which are situated near the skin, provoke two other mechanisms for regulating the body's temperature:

- *Vasoconstriction*: It is expressed in reduction of the blood flow close to the skin. The result is fast cooling of the extremities, which, in the case of outdoor conditions, can lead to limb injuries. The concentration of blood flow inside the core body ensures the necessary temperature for the important internal organs.
- *Shivering*: It is based on an increase in the internal heat generation by stimulating muscles and provoking contraction.

The temperature regulation is governed by warm and cold impulses (Kosaka et al. 2004). The number of impulses sent to the brain is dependent on how quickly the temperature changes. When the body is totally relaxed and in thermophysiological comfort, no impulses for warm or cold are sent to the brain. Therefore, the conditions for thermophysiological comfort from the point of view of the body temperature regulation may be defined (Kosaka et al. 2004) as follows:

- The combination of core body temperature and skin temperature must provide a sensation of thermal neutrality.
- The heat produced by metabolism must be equal to the heat released by the body: body heat storage tends to be zero.

The processes of heat production and heat losses between the human body (skin) and the environment are discussed in detail in the monograph of Havenith (2003). The processes are complicated, however, if a textile layer (or layer system) exists between them (Angelova 2008).

Clothing is also a potential cause of thermal discomfort. Initial studies on the interaction of garments and the human body have been conducted for military purposes. The reason is that the individuals adapt to the environment, in which they live and work, but they can usually add a layer of clothing if they feel cold, or remove a clothing item if they feel hot. In the case of uniforms or protective clothing, this option is disabled. The problem is valid for workers clothing as well, so even in an office building the question of the effect of the insulating properties of clothing on the thermophysiological comfort of the inhabitants can be very serious.

4.2 Mechanisms of Heat Loss and Heat Generation of the Human Body

4.2.1 Mechanisms of Heat Loss

The heat losses from the body in the presence of a textile layer(s) are associated with the following heat transfer processes (Angelova 2007a):

- *Conduction*: This is the mechanism of heat losses through contact with a cooler object and transfer of heat to the cooler object, that is, the contact between the skin and a cooler textile layer or the contact between two adjacent layers of fabrics (the outer layer is colder than the inner). Heat losses by conduction increase when the colder surface is wet. In general, conduction accounts for 2%–3% of total heat losses in dry conditions.

  If the textile layer between the body and the environment is wet, the heat losses increase five times. Moreover, if the body is immersed in cold water, the rate of heat losses is 25 times faster, as the water conductivity is around 25 times higher than that of the air (Nielsen 1978; Tarlochan and Ramesh 2005).

- *Convection*: Heat losses from the body by convection are associated with the removal of the layer of warm air between the body and the textile layers, that is, losses by convection heat transfer through the air layer between the skin surface and the closest textile layer and losses by convection heat transfer through the air layers between different textile layers (from inner to outer layers).

  The rate of convection heat losses depends on the air velocity. The air velocity is very small in the indoor environment and occurs mainly due to body movement: the so-called pumping effect appears when the textile layer wraps round the body (see Chapter 3). In the external environment, heat losses by convection, however, may be particularly important because of the presence of forced convection (wind).

- *Radiation*: It is related with the heat losses from the body to a colder environment due to a temperature difference:
  - Radiation heat losses from the uncovered parts of the body to the surrounded air
  - Radiation heat losses from the outer textile layer to the surrounded air

However, radiation heat losses from the uncovered parts of the body (hands, face, head) are much higher than from the outer textile (clothing) layer.

- *Evaporation*: This is the mechanism of heat losses in hot environment in the form of water vapor when the body uses heat to evaporate moisture from skin surface (sweat). Two types of heat losses appear:
  - Heat losses by latent heat transfer (evaporation of water) from uncovered parts of the body.
  - Heat losses by latent heat transfer from the surface of the outer textile layer.
- *Perspiration*: This is the mechanism of heat losses from lungs, related to warming of the inhaled colder air, which is then exhaled.

Finally, the factors that affect the heat losses from the human body in its interaction with the environment and textiles may be summarized as follows:

- Air temperature, air velocity, wetness
- Area of skin surface which is not covered (not protected by textiles)
- Presence of contact and size of the contact area with surfaces or water

4.2.2 Mechanisms of Heat Generation

In order to counterbalance the heat losses, the body must produce an equal amount of heat so as to stay active despite the temperature of the environment.

The mechanisms used by the body to produce heat are as follows:

- *Metabolism*: Biochemical reactions appear in the body and the heat is produced as a by-product.
- *Physical activity*: Muscles generate heat during physical work. Most of body's heat is produced by this mechanism.
- *Shivering*: Heat is generated via uncontrolled quivering of the muscles in cold environment that increases the heat production. The effect of this mechanism is limited to a few hours because of depletion of glycogen stores (the muscle *fuel*) and the appearance of fatigue.

Concerning the factors that influence the heat production in the body, two groups can be distinguished.

- Factors important for body's heat production:
  - Food intake
  - Metabolism, determined mainly by the activity

- Glycogen store
- Fluid balance
- Factors important for body's heat retention (and tolerance to cold environment):
  - Thermal insulation of the textiles/clothing
  - Size and shape of the body: body surface to body volume ratio
  - Layer of fat under the skin
  - Sorption abilities of the textiles
  - Temperature of the environment

Thermophysiological comfort, as part of the overall comfort of the person, is assessed by various sensory signals that are processed by the brain. Two main sensations can be distinguished: *temperature* and *humidity* (Table 4.1).

The thermal discomfort associated with the presence of a sensation for moisture is much more discussed than the discomfort preconditioned by cold indoor environment. Clothing or textiles can further increase the discomfort due to the moisture that appears because of the imbalance between the generated (or absorbed) heat by the body and heat losses. The following aspects related to the sensation of moisture should be taken into account:

- Water vapor or sweat is transported from the moist skin surface to the nearest textile layer (which can be part of clothing, furniture fabric, bedding, etc.). As a result, the friction coefficient *textile–skin* is changed, which provokes as a rule an unpleasant sensation in contact.
- Being wet due to the convective moisture transport or the diffusive transfer to the environment (see also Chapter 2), the textile layer cools fast and provokes unpleasant sensation of wetness in case of direct contact with skin (cold sensation).

TABLE 4.1

Sensations of the Person Related to Thermophysiological Comfort

Sensation	Integral Result of	Sensation	Sensory Signal From
Temperature sensation	Sensation for temperature	Cold/warm	Skin
	Speed of temperature changes	Speed of cooling/warming	
Moisture sensation	Sensation of temperature changes	Increased speed of cooling	No special sensors
	Tactile sensors	Wet skin, drops of sweat	

4.3 Impact of Textiles in the Indoor Environment on the Human Body

4.3.1 Effect of Chairs

When the chair is covered with a textile layer, the clothing insulation I_{cl} of a sitting person increases by up to 0.15 clo depending on the contact area between the chair and body. This value has to be added to the total clothing insulation, following Equation 3.13 (Chapter 3). A desk chair, for example, with a body contact area of 2700 cm^2, has a clothing insulation of 0.1 clo (McCullough et al. 1994) (Table 4.2). This amount should be added to the insulation of the standing clothing ensemble to obtain the insulation of the ensemble *clothing–chair* when the person is sitting.

4.3.2 Effect of Bedding Textiles

The body's thermoregulation is less active during sleep than in waking conditions, for example. Therefore, the bedding textiles should guarantee the formation of an additional *microenvironment* around the body, different than the environment in the room. Tests, performed in different conditions, have identified that the thermo-neutral zone of the additional microenvironment is around 30°C (Muzet et al. 1984). It was found that when the ambient temperature changed between 19°C and 22°C, the temperature of the microenvironment remained at 29.6°C. The drop of the ambient temperature to 13°C provoked decrease in the temperature of the microenvironment to 26.1°C.

The microenvironment around the resting human body depends on the following:

- The temperature of the environment (being it indoor or outdoor)
- The design and application of the bedding item (blanket, duvet, etc.)
- The clothing of the person

The heat losses from the microenvironment around the resting human body are related with the following basic processes:

TABLE 4.2

Clo-Values for Different Types of Chairs

Type of the Chair	Clo-Value
Wooden or metal chair with solid seats and backs	0.0
Fabric covered, cushioned	0.10–0.16
Arm chair	0.20–0.32

Source: McCullough, E.A. et al., *ASHRAE Tran.*, 100, 795–802, 1994.

- The leakage of air from the microenvironment to the indoor environment through the upper layers of bedding
- The conductive heat transfer through the mattress

The bedding textiles, whose main purpose is to create a comfortable microenvironment around the human body, are both traditional and electric blankets, and duvets.

4.3.2.1 Duvets

The duvet is a soft bag, usually cotton one, traditionally filled with down or feathers in different proportion, hay, straw, rags, and so on, used as a blanket. Nowadays, a duvet can be filled with natural fibers (cotton, wool, silk), down, and feathers or with man-made fibers or webs (mainly polyester).

The main advantage of duvets is that they are usually used as a single-layer bedding textile, as they ensure the necessary temperature of the microenvironment around the resting body without the need to be combined with other bedding items. At the same time, Holland et al. (1999) have found that duvets have the tendency to overheat the body since they allow little heat exchange between the indoor air and the microenvironment. Therefore, the authors suggested the use of duvets only in indoor environment with proper thermal conditions (lower temperatures).

From the point of view of thermophysiological comfort, the duvets in the indoor environment have to meet the following requirements:

- To obtain insulating, moisture absorbing, and temperature compensating effects.
- To adapt to body shape. Due to the frequent change of the sleeping position and thickness of duvets, the ventilation effect leads to decrease in the temperature of the microenvironment (Holland et al. 1999).

4.3.2.2 Common Blankets

Common blankets are one-layer or two-layered textiles from different materials. Wool fibers are able to buffer temperature extremes and changes in relative humidity of the environment. Wool blankets also transport more sweat away from skin as compared to cotton blankets and especially to blankets from man-made fibers. Figure 4.1 shows comparative results for the maximum weight of absorbed water in relation to the dry weight of three types of blankets.

Umbach (1986) has found that wool blankets provided 8%–20% higher thermal insulation and absorbed 50% more sweat, confirming the theoretical expectations. What's more, under an acrylic-cotton blanket, 75% of subjects felt uncomfortably hot and 88% felt clammy, while the comparable figures for the wool blanket were only 38% and 50%.

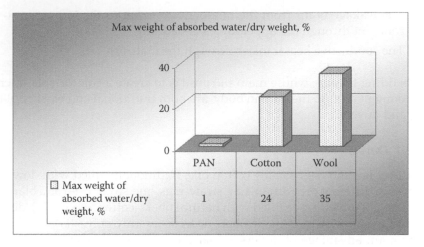

FIGURE 4.1
Maximum weight of absorbed water in relation to the dry weight of three types of blankets.

4.3.2.3 Electric Blankets

Electric blankets create microenvironment, which is not related to the dynamic characteristics that adjust in accordance with the change of the temperature of the indoor environment. Mizuno et al. (2005) have found that the use of electric blankets under low ambient temperatures of 3°C and relative humidity of 50%–80% is beneficial for sleep. Lower extremities are the zones for the best heating. However, electric blankets can cause thermal stress in case of higher ambient temperatures. In such cases, the blanket has to be switched off during the night.

4.3.3 Some Solutions for Improving the Effect of Textiles on Human Thermophysiological Comfort

Several solutions for improving the role of fabrics and clothing as an insulation barrier can be found in their design and properties. Of course, this concerns only *common* textiles, which are used in everyday life and have no special treatments for low permeability, high or low temperatures resistance, and so on.

The general rule for having a person in thermal neutrality regarding the textiles, he is surrounded, is the moisture to be kept away from the skin. This can be achieved by the following:

• Design and application of two-layered fabrics. Low absorption, high conductivity fibers are used from the inside, close to the skin layer, while high absorption fibers are applied in the manufacturing of the outer layer.

- Use of natural fibers like cotton, wool, silk, and flax in the mesostructure, so as to assure very high moisture absorption abilities.
- Application of special weaving and knitting techniques for the production of macrostructures, which can retain moisture between the yarns (towelling is a typical example for this).
- Use of chemical fibers with a hollow cross section in the mesostructure again for trapping the moisture inside.
- Use of elastomeric threads in textile macrostructures for the production of tight fitting clothes, so as to reduce the friction between the skin and the closer to the skin textile layer. On microstructural level, this has to be combined with fibers with high conductivity or moisture absorption abilities.
- Use of loose fitting clothes, which can reduce significantly the thermal discomfort due to high speed of cooling when high moisture absorption fibers are used on microstructural level.

Obviously, when the moisture is transferred away from the skin by using high conductivity chemical fibers, possible skin problems like irritant dermatitis, allergic contact dermatitis, and immediate-type reactions have to be taken into consideration. Several studies have shown that people are much more sensitive to the use of chemical fibers in the textiles than to the presence of protein fibers (Havenith 2003). The same is valid for fabric softeners, used frequently to reduce the *skin–fabric* friction, although they are *weaker* allergens than the fibers themselves.

4.4 Summary

The general principles of the interaction between textiles/clothing and the human body for ensuring thermophysiological comfort were presented.

The mechanisms of heat loss and heat generation by the human body in the case of presence of clothing were analyzed.

The ways for impact of the textiles on the human body in the indoor environment were discussed. The necessity for creation of a microenvironment around the human body at rest (sleep) was described and analyzed.

Section II

Experimental Study of Woven Textiles Used in the Indoor Environment

5

General Considerations and Methodology of the Experimental Study

5.1 Aims and Scope of the Experimental Study

The experimental study in Section II is dedicated to *woven textiles* with various applications. The choice of a woven macrostructure is predetermined by its basic characteristics:

- It is in possession of all levels of the textiles hierarchical structure: micro-, meso-, and macrostructure (which differs from the nonwoven textiles).
- The woven fabrics have relatively well-arranged macrostructure, which gives way to systematic geometric description and implies a correct analysis and assessment of the characteristics of the macrostructure and its behavior in relation with thermophysiological comfort.

The research in Section II is subject to the following general requirements:

- To evaluate the impact of specific geometric, structural, and mass characteristics of the meso- and macrostructure on the properties of woven textiles, related to thermophysiological comfort: transport of fluids and heat.
- To study woven textiles with various applications: for clothing, surgical clothes and linen, furniture fabrics, fabrics for packing.
- To derive analytical equations that can be used out of the scope of the particular experiment.
- To create a database with structural and functional parameters that can be used in the computer simulation of heat and mass transfer properties in Section III.

In Section II of the book, properties of the woven textiles, related with transport of fluids and heat, are investigated as a function of characteristics of their meso- and macrostructure as follows:

- Experimental investigation of the mesostructure of woven textiles (Chapter 6)
- Experimental investigation of the macrostructure of woven textiles for clothing and linen (Chapter 7)
- Experimental investigation of the macrostructure of woven textiles for surgical clothes and medical linen (Chapter 8)
- Experimental investigation of the macrostructure of woven textiles for furniture (Chapter 9)
- Experimental investigation of the macrostructure of woven textiles for packing (Chapter 10)
- Experimental investigation of textile macrostructures in the indoor environment (Chapter 11)

5.2 General Description of the Study

The general layout of the experimental study as it concerns the methodology of the experiment, the experimental conditions, and the used appliances is presented. In the following chapters, only particularities of the study are commented (if any).

5.2.1 Experimental Investigation of the Mesostructure

5.2.1.1 Linear Density

The linear density of the threads was measured in accordance with EN ISO 1973 (1999). The length of the hanks was 100 m with a constant tension of 0.5 cN/tex. The weight of the hanks was measured on a digital scale. The linear density was determined using the expression:

$$Tt = \frac{m_g}{L_g}, \text{tex} \tag{5.1}$$

where:
m_g is the weight of the hank, g
L_g is the length of the thread in the hank, km

The following procedure was applied for measurement of the linear density of threads from woven samples: from each sample, 10 warp and 10 weft threads were unraveled. The length and the weight of each thread were measured and Equation 5.1 was applied. The means of the 10 measurements were calculated for both warp and weft threads.

5.2.1.2 Twist

The twist was determined in accordance with EN ISO 2061 (2010). The length of the sample was 25 mm with preliminary tension of 0.5 cN/tex. The same methodology was applied for warp and weft threads from woven samples. Ten single measurements were performed for calculation of each mean variable.

5.2.1.3 Width of Ribbons

A nondestructive analysis and microscopy were used for the measurements. The width of 10 warp and 10 weft ribbons was measured at 10 different points each and the mean width was calculated.

5.2.1.4 Tenacity and Elongation at Break

They were measured by using electronic dynamometer according to the requirements of EN ISO 2062 (2010). The following test conditions were applied: preload of 0.5 cN/tex, 500 mm length of the sample, and breaking time 20 ± 3 s. The elongations at break E, %, was measured directly, while the tenacity R, cN/tex, was calculated on the basis of the measurement of the absolute strength P:

$$R = \frac{P}{Tt} \qquad (5.2)$$

5.2.2 Experimental Investigation of the Macrostructure

5.2.2.1 Warp and Weft Density

ISO 7211-2 (1984) was used and the number of threads per 10 cm was calculated to obtain warp P_{wa} and weft P_{wf} density, respectively.

5.2.2.2 Thickness

The thickness δ was determined following EN ISO 5084 (2002). A sample with an area of 25 cm^2 was used, with a pressure of 0.5 kPa.

5.2.2.3 Yarn Crimp

ISO 7211-3 (1984) was followed and 10 threads in warp and 10 threads in weft direction were measured for each of the samples. The length of the sample was 200 mm. The length of the straight yarns L_n was measured, and yarn crimp in warp and weft directions (a_{wa} and a_{wf}) was calculated in accordance with the equation

$$a_{wa, wf} = \frac{L_n - L_t}{L_t} 100, \%$$ (5.3)

where:
   L_t is the length of the thread in the woven sample, mm

5.2.2.4 Fabric Weight

The ISO 3801 (1977) standard was used. The sample area was 100 cm², and the fabric weight was determined using the equation

$$m_S = \frac{m_o}{L_o \cdot B_o}, \text{ g/m}^2$$ (5.4)

where:
   m_o is the mass of the sample, g
   L_o is the length
   B_o is the width of the sample, m

5.2.2.5 Warp and Weft Cover Factor

The cover factor shows how much of the fabric in the direction of warp threads is filled in by the weft threads and vice versa. The warp cover factor E_{wa} was determined by Equation 5.5, and the weft cover factor E_{wf} by Equation 5.6:

$$E_{wa} = k_f P_{wa} \sqrt{\frac{Tt_{wa}}{1000}}, \%$$ (5.5)

$$E_{wf} = k_f P_{wf} \sqrt{\frac{Tt_{wf}}{1000}}, \%$$ (5.6)

where:
   Tt_{wa} is the linear density of the warp threads
   Tt_{wf} is the linear density of the weft threads, tex
   k_f is a coefficient that depends on the fiber type

5.2.2.6 Fabric Cover Factor

The fabric cover factor E_s describes the area of the fabric, which is filled in with threads from both systems (warp and weft threads). It was determined using Equation 5.7:

$$E_s = E_{wa} + E_{wf} - 0.01 v_{wa} v_{wf}, \%$$ (5.7)

5.2.2.7 Fabric Areal Porosity

It reflects the area of the inter-yarn pores compared to the area of the fabric. It was calculated as

$$V_s = 100 - E_s, \%$$ (5.8)

5.2.2.8 Air Permeability

EN ISO 9237 (1999) was applied during the measurement. The area of the sample was 10 cm² and 100 Pa pressure difference was applied from both sides of the sample. The air-permeability coefficient was calculated following Equation 5.9:

$$B_p = \frac{Q_a}{360 \cdot A}, \ \text{m/s}$$ (5.9)

where:
Q_a is the air flow rate, dm³/h
A is the area of the sample, cm²

5.2.2.9 Thermal Resistance

Guarded heat flow meter was used in accordance with ASTM E1530 (1999). The area of the sample was 50 mm with 100 Pa pressure between the two hot plates. Three measurements were performed for each sample and the mean of the thermal resistance R_t was determined. The thermal insulation I_{cl} of the samples was calculated following Equation 3.9 (Chapter 3).

5.2.2.10 Liquid Transfer

The AATCC 195 (2012) standard was used, and the characteristics were incorporated in moisture management tester (SDL Atlas 2010), namely, WTT (wetting time top), s; WTB (wetting time bottom), s; TAR (top absorption rate), %/s; BAR (bottom absorption rate), %/s; TSS (top spreading speed), mm/s; BSS (bottom spreading speed), mm/s; R_{index} (accumulative one-way transport index), mm²/s; and OMMC (overall moisture management capacity). Each of the samples was 80 × 80 mm and five measurements were performed for each macrostructure.

5.3 Calculation of the Relative Error

The relative error allows the comparison between theoretical and experimental values, and it is determined by

$$p_o = \frac{(x_e - x_t)}{x_e} 100, \%$$ (5.10)

where:

x_t is the theoretical value

x_e is the experimental value (i.e., the mean from several measurements)

5.4 Summary

The aim, scope, and general considerations of the experimental study on the properties of woven structures with respect to their thermophysiological comfort, presented in Section II, were described.

The general conditions and methods for experimental investigation of the meso- and macrostructure of woven textiles were presented.

6

Experimental Investigation
of the Mesostructure

The main aim of the experiment, presented in this chapter, was to examine the impact of the mesostructure on the transfer of air, heat, and moisture through woven macrostructures, looking for structural parameter of the constituent yarns that would have statistically proven influence on the heat and mass transfer through the macrostructure.

6.1 State of the Art of the Problem: Impact of the Mesostructure on the Transfer of Heat and Fluids

Studying the woven textiles properties related to thermophysiological comfort requires detailed knowledge of meso- and macrostructure of the fabric. This is necessary also in the case of production of fabrics with required functional properties for a particular application (i.e., air bags, precise woven structures for medical use) (Angelova 2011).

The hierarchical structure of textiles, and in particular of the woven textiles, makes the problem of the evaluation of their functional properties, associated with the transport of fluids and heat, quite complex. The fluid motion in through-thickness direction is performed not only through the pores in the macrostructure but also through the pores in the mesostructure, between the individual fibers in the yarns or monofilaments in polyfilament threads (Gebart 1992; Dubrovski and Sujica 1995; Wong 2006; Angelova 2012). The microstructure may also affect the transport of air as the moisture content of the fibers can significantly alter the air permeability of the macrostructure (Gibson et al. 1999).

According to the analysis in Chapter 2, the air between the fibers or monofilaments in the mesostructure is a reason for the

- Presence of a proportion of the total air permeability of the woven fabrics, which is due to the voids in mesostructural level.
- Heat transfer by convection in the mesostructure.

There are two main characteristics of the mesostructure that can affect its permeability: the linear density of the yarns Tt and their twist Ts. They are related to the size of the cross section of the yarns (diameter) and their compactness (Konova and Angelova 2013).

Twisting is the process which removes to a large extent the air detained between the fibers in the slivers or rovings. The higher the twist is, the smaller the diameter of the yarn is and the yarn structure is more compact. Therefore, it is reasonable to expect that the conventional ring spun yarns have greater diameter than the compact ring yarns.

Rotor yarns are an exception: though they are spun with higher twists, they are less compact than the ring spun yarns (Kullman et al. 1981; Paek 1995; Basal and Oxenham 2006; Angelova 2010a). The reasons are in both phases of the morphology of the spinning process. The discretization of the sliver to individual fibers and the join to the open end provoke the specific distribution of the twist in the rotor yarns, more twists in the core and less on the periphery of the yarn. Thus, bulkiness of rotor yarn reaches 125% than that of analogous ring yarn.

In the case of compact ring spun yarns, the absence of a spinning triangle in the second phase of the morphology of the spinning process results in yarns with a reduced diameter and hairiness. The increased twist angle causes a greater diameter reduction with the augmentation of twist in comparison with conventional ring yarns (Basal and Oxenham 2006).

In light of this analysis, it seems logical the conclusion of Kullman et al. (1981) that due to the lower density of the rotor yarns, the fabrics from them have higher air permeability. Therefore, data for the mesostructural characteristics are necessary for a proper evaluation of the behavior of the woven macrostructures and for more accurate analysis of the effect of the hierarchical structure in case of simulation of permeability or heat transfer processes. This requirement is even stronger if the numerical simulation of the heat transfer process is performed on a micro- or mesostructural level.

In fact, a number of studies have dealt with the air permeability of woven textiles, mainly in relation to their porosity. Many of them were oriented toward the investigation of the impact of parameters of the macrostructure on the transmission of air, associated with both air permeability and convective heat transfer. No systematic studies have been performed, however, on the effect of the mesostructure on these processes.

In his work Clayton (1935) showed that the air permeability of woven structures decreases with the increment of the linear density of the constituent yarns. He found out that the higher twist values of warp and weft threads caused an increase in the measured values of air permeability, but he did not discuss the results in connection with the changed size of the pores in the woven macrostructure. Lamb and Constanza (1979) concluded that the higher bending of threads from one of the systems around the other (related to the Novikov's phase of construction) increased the air permeability. The authors explained their findings with the higher elasticity of the macrostructure. Epps and Song (1992) established a positive correlation (linear correlation coefficient rxy > 0) between the twist of yarns and the ability of the fabric to retain air. At the same time, they claimed that there was no connection between the linear density of the threads and the air permeability, which did not coincide with the research findings of Clayton (1935).

The impact of the mesostructure was indirectly evaluated in studies dedicated to investigation of the porosity of woven structures. However, quantitative dependences between measurable characteristics of the yarns and the experimentally evaluated air permeability were not established (Kulichenko and Van Langenhove 1992; Epps and Leonas 1997; Dubrovski 2000; Cay et al. 2004; Xu and Wang 2005; Ogulata 2006).

The absence of a systematic approach in the investigation of the effect of the mesostructure on the textiles air permeability was discussed in the work of Singh and Nigam (2013). Their conclusion was made on the basis of an extensive literature overview. Shortly later, the work of Ishtiaque et al. (2014), which explored the transmission of air and heat through woven and knitted macrostructures, was published.

Unlike research works focused on the influence of the mesostructure, there are a series of studies on textiles based on the application of analytical models. Analytical models were applied for the investigation of the transport processes through porous media, and in particular trough textiles. They were used mainly for studies on nonwoven macrostructures, but also on yarns in woven and knitted macrostructures. The most popular models were those of the Kozeny (1927) and Carman (1956), but the models of Davies (1952), Piekaar (1967), and others were also applied. It should be noted that the model of Carman (or Kozeny–Carman) is used only for fluids where the Darcy's law is valid (with Reynolds number Re < 1) and the fluid moves in a plane parallel to the textile surface.

Gebart (1992) proposed an analytical model used for calculation of the fluid transfer in both parallel and perpendicular direction of a polyfilament. He considered two ways of arrangement of the monofilaments in the complex thread—square and hexagonal—and performed a 2D simulation of transfer of a Newtonian fluid in a composite material.

Kulichenko and Langenhove (1992) also developed an analytical model that gave good results only for prediction of flow through the pores of the macrostructure. A combination of this model and the model of Gebart (1992) was used in Saldaeva (2010) for calculation of the air permeability of woven macrostructures. Twelve samples were examined for determination of the transmission of air and oil in transverse direction of the macrostructure. Essential part of the experimental results, however, differed from the calculations. Saldaeva (2010) investigated also the influence of pressure on the textiles permeability and found out that the relationship is nonlinear, which has been shown in previous studies as well (Douglas and Huiping 1992; Wang and Liu 2004).

6.2 Properties of the Yarns

Six different cotton ring spun yarns were used. Their linear density changed from 20 to 50 tex, and their mean twist decreased from 1045 to 550 m⁻¹, respectively. The yarns were used for manufacturing of eight woven samples. All yarns were tested for determination of structural and mechanical properties: twist, strength, tenacity, elongation at break, and so on.

The results for the twist are summarized in Table 6.1: mean value, twist irregularity (expressed by the variation coefficient), and the relative standard error. Twist factor for each yarn was also calculated.

It is logical to expect that the yarns with higher linear density (i.e., sample 6) will allow more air flow to pass through them, which are part of the woven macrostructure, than finer yarns (i.e., samples 1 and 2). The reason is the higher twist used for the production of yarns with lower linear density, in order to ensure the necessary levels of strength and elongation at break. At the same time, studies on air permeability have shown that macrostructures with higher surface density, produced from yarns with a higher linear density, have lower values of the coefficient of air permeability (Dhingra and Postle 1977; Epps and Song 1992; Epps 1996, etc.).

The results for the yarns tenacity and elongation at break are shown in Table 6.2.

6.3 Manufacturing of Identical Woven Structures from Different Yarns

Epps and Song (1992) evaluated the impact of linear density, twist, and twist direction on permeability and heat transfer of woven fabric, investigating three samples of plain weave polyester fabrics. The studied samples had completely different structural and mass characteristics: warp and weft

TABLE 6.1

Basic Properties of the Ring Spun Yarns

Sample	Linear Density, tex	Twist, m⁻¹	Twist Factor	Twist Irregularity, %	Relative Standard Error, %
1	20	1045	148	5.07	2.4
2	25	935	148	3.71	1.7
3	28	785	131	4.19	1.9
4	30	775	135	3.67	1.7
5	36	760	145	3.95	1.9
6	50	550	162	5.21	2.4

TABLE 6.2

Yarn Tenacity and Elongation at Break

Sample	Tenacity R, cN/tex	Tenacity Variation Coefficient V_R,%	Relative Standard Error, %	Elongation at Break ε, %	Elongation Variation Coefficient Vε,%	Relative Standard Error, %
1	10.52	10.15	2.1	4.97	13.87	2.9
2	9.49	12.30	2.6	5.02	13.12	2.7
3	11.88	9.75	2.0	5.81	8.65	1.8
4	12.02	9.93	2.0	6.15	9.67	2.0
5	11.09	11.12	2.3	6.16	11.44	2.4
6	11.07	7.87	1.6	6.51	7.52	1.6

densities, areal density, thickness, and so on. The authors commented the lack of identical parameters of the macrostructures, but made analyzes and drawn conclusions, assuming that all fabrics were of *medium weight*.

This compromise may be accepted, indeed, but it requires at least an increase in the number of tested samples. In fact, Epps and Song (1992) touched the heart of the problem, but did not discuss it.

Other researchers, who claimed to investigate the properties of fabrics with different mesostructure (Singh and Nigam 2013; Ishtiaque et al. 2014), didn't do it either. The essence of the problem, related with the correct analysis of the effect of mesostructure on functional properties of the macrostructure, is that fabrics with the same parameters should be produced (or simulated) from yarns with different characteristics. This leads to specific difficulties defined as the following proposition: *The production of identical woven macrostructures (i.e., with the same geometric, structural and mass parameters), while changing the characteristics of the mesostructure—linear density and twist, is impossible.*

In defense of the proposition, two statements, related to the effect of the yarns linear density on characteristics of the macrostructure, are defined and proved. The twist is implicitly linked to the dimensions of the yarns cross section: the greater the twist is, the more the size of the cross-sectional area decreases and the thread becomes more compact. In this sense, the increase in the twist would have the same effect on the size of the cross section as the lessening of the linear density and vice versa.

- *Statement one*: The use of yarns of different linear density Tt, while maintaining constant warp and weft densities, results in a woven macrostructures with different characteristics.

Theoretically, there is no technological problem for the production of two woven fabrics with the same densities: the number of threads per cm in both directions of the fabric is set to be equal. This requires the respective calculations and settings in both warping and mechanism for take-up of the fabric on the weaving machine to be done.

To prove statement one, a set of 19 woven fabrics with equal warp and weft densities $P_{wa} = P_{wf} = 10$ threads/cm was virtually designed. The linear density of the warp threads was equal to the linear density of weft threads and varied in the range 10–100 tex with a step of 5 tex. The distance between warp and weft threads, and pore size, was determined.

Figure 6.1 shows the results from the calculations. It is obvious that with the increase in the linear density of the constituent threads, the pore size decreases in the direction of both warp and weft threads. The pore area decreases as well, which will inevitably affect the porosity of the woven macrostructure. Thus, the flow rate through a single pore of the macrostructure of fine yarns would be greater, regardless the reduced flow rate through the yarn itself.

- *Statement two*: The use of yarns of different linear density Tt, while maintaining constant the distance between warp and weft threads, results in a woven macrostructures with different characteristics.

Uniform spacing between the warp threads of the two macrostructures can be controlled by the mechanisms of the weaving machine and the uniform distance between the weft threads—through the use of negative take-up mechanism, for example.

To confirm statement two, a second set of 19 woven structures was virtually designed, so as to maintain equal distance between warp and weft threads: 0.314 mm in both directions. Warp and weft threads were again of the same linear density, which varied in the range 10–100 tex with a step of 5 tex. The change of warp and weft densities was determined together with the number of pores in 1 dm² of the fabric. Figure 6.2 summarizes the results from the calculations.

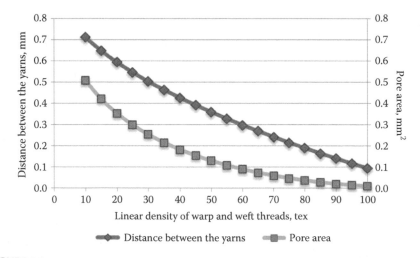

FIGURE 6.1
Distance between threads and pore area for different yarns' linear densities and constant warp and weft density of the macrostructure.

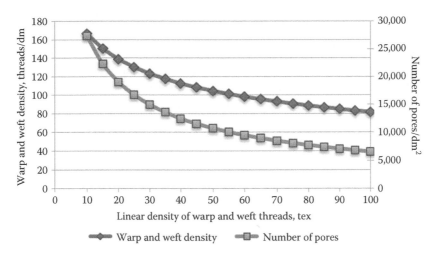

FIGURE 6.2
Warp and weft densities and number of pores in dm^2 for different yarns' linear densities and constant of distance between the threads.

Obviously, by maintaining the same distance between threads (and the same area of the pores between them), the decrease in the yarns' linear density leads to the production of a macrostructure with higher warp and weft densities. This not only increases the number of pores per unit area of the macrostructure but also provokes higher resistance of the yarns to fluid flows in through-thickness direction.

In conclusion, the change of the linear density or twist of the mesostructure leads to weaving of new macrostructures, which cannot be treated as identical, having different characteristics.

Thus, the single-impact influence of the linear density and/or the twist of the yarns on the transport processes through the macrostructure cannot be independently evaluated. Obviously, the approach used in the literature, although connected with flaws, is the only possible way.

In light of the formulated and proven proposition, an attempt was made in this chapter to evaluate the effect of mesostructural characteristics on the functional properties of woven macrostructures, namely, heat and mass transfer processes in through-thickness direction, ignoring the existence of other differences between the investigated macrostructures and using the methods of mathematical statistics.

The same approach was applied in the work of Singh and Nigam (2013) based on detailed state-of-the-art analysis. The authors concluded that there wasn't so far a systematic study on the effect of the yarn properties (in particular ring yarns) on the comfort of textiles. Two types of ring yarns, conventional (carded and combed) and compact, were used for the production of 12 woven macrostructures for shirts. Despite the differences in thicknesses, weft densities, and fabric weight, the authors examined the effect of the mesostructure on the transfer of heat, air, and moisture through the macrostructures.

A similar approach was used by Ishtiaque et al. (2014). They researched the impact of three values of the yarn twist on the transfer of heat and fluids through three woven and three knitted macrostructures. The woven samples were unbalanced in terms of densities, but the threads in the mesostructure were of the same linear density (20 tex). An advantage of the study was the simultaneous assessment of woven and knitted macrostructures, but a similarity to the work of Epps and Song (1992) shortcoming appeared: the conclusions were made on the basis of three samples only.

In the present study, eight woven macrostructures were designed from the yarns, described in Table 6.1. Sulzer projectile machine was used for the manufacturing of the samples. Unbalanced fabrics in terms of warp and weft densities were produced, but the threads used in both warp and weft direction had equal linear densities. The fabric weight of the macrostructures was from light to medium and plain weave was applied in order to avoid the influence of different number of interlacing in case of using various patterns.

Tables 6.3 and 6.4 summarize the main characteristics of both the gray and washed cotton fabrics, respectively. The effect of the desizing process on warp and weft densities, thickness, and fabric weight was analyzed. Similar to the approach used in the literature (Havlova 2013; Singh and Nigam 2013; Ishtiaque et al. 2014), the analyzes were based on the original twists and linear densities of yarns used in the production of the gray fabrics. In his work Saville (1999) pointed out that it would be better to take into account the initial linear density of the yarns used to produce the raw fabric because of the possibility the yarns in the finished fabric to have either larger mass per unit length or a smaller one.

Figure 6.3 shows microscopic photographs (4× enlargement) of the two of the macrostructures. The textiles, in accordance with the fabric weight, were appropriate for the production of clothing and bedding.

TABLE 6.3

Characteristics of the Gray Fabrics

Sample	Fabric Weight, g/m²	Warp Density, threads/dm	Weft Density, threads/dm	Linear Density, tex		Weave	Thickness, mm
				Warp	Weft		
1	62	136	148	20	20	Plain	0.40
2	84	168	100	28	28	Plain	0.42
3	123	272	268	20	20	Plain	0.36
4	138	264	238	25	25	Plain	0.40
5	153	223	230	30	30	Plain	0.42
6	187	233	234	36	36	Plain	0.44
7	195	237	243	36	36	Plain	0.44
8	233	258	150	50	50	Plain	0.61

TABLE 6.4

Characteristics of the Finished Fabrics

Sample	Fabric Weight, g/m²	Warp Density, threads/dm	Weft Density, threads/dm	Thickness, mm
1	63	142	166	0.43
2	89	176	124	0.45
3	133	284	294	0.39
4	138	270	266	0.41
5	157	234	254	0.47
6	201	258	238	0.49
7	197	252	260	0.48
8	235	267	166	0.62

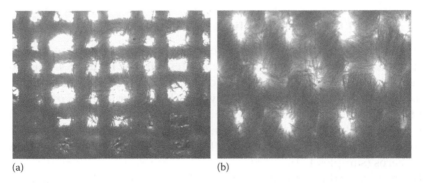

(a)   (b)

FIGURE 6.3

Microscopic pictures of (a) sample 3 (133 g/m², 20 tex) and (b) sample 8 (235 g/m², 50 tex).

6.4 Analysis of the Effect of the Mesostructure on the Air Permeability of the Macrostructure

Table 6.5 summarizes the mean values from the measurements of the air flow rate in through-thickness direction of the tested macrostructures and the calculated air permeability coefficient for both gray and finished fabrics. The results of the experiment showed that the finished fabrics had lower air permeability, which was expected as the wet processes led to relaxation of the threads after weaving. The fabric weight and thickness increased due to the shrink of the samples in both longitudinal and transverse direction. All this resulted in increased warp and weft densities and reduced pore size of the finished fabrics, which influenced their air permeability. The statistical analysis of the results for the air permeability coefficient indicated that the finishing treatment had proven impact on the flow rate through the samples (0.05 significance level).

TABLE 6.5

Air Permeability of the Macrostructures

	Air Flow Rate Q_a, m³/s		Air Permeability Coefficient B_p, m/s	
Sample	Gray Fabrics	Finished Fabrics	Gray Fabrics	Finished Fabrics
1	0.00276	0.00260	1.38	1.30
2	0.00270	0.00250	1.35	1.25
3	0.00166	0.00158	0.83	0.79
4	0.00128	0.00120	0.64	0.60
5	0.00152	0.00124	0.76	0.62
6	0.00098	0.00068	0.49	0.34
7	0.00116	0.00050	0.58	0.25
8	0.00184	0.00046	0.92	0.23

6.4.1 Effect of the Linear Density

The relationship between the linear density of the mesostructure Tt and the air permeability of the macrostructure is presented in Figure 6.4. The established dependence showed diminution of the air permeability coefficient B_p of the fabrics, woven with coarse yarns. However, the use of warp and weft threads wit Tt = 50 tex (Sample 6) resulted in an increase of B_p. This was probably due to a greater proportion of the flow through the pores on a mesostructural level (pores between the fibers), compared with yarns with finer yarns (samples 1–4).

The reason for this assumption is again the technological fact that the finer yarns are produced with higher twists so as to reach at least the minimum requirements for strength and elongation at break. Thus, the space between the fibers in the fine yarn decreases (as compared to a yarn with higher values of Tt) and the air flow passes mainly through the pores of the macrostructure. The regression equation for the relationship between the linear density

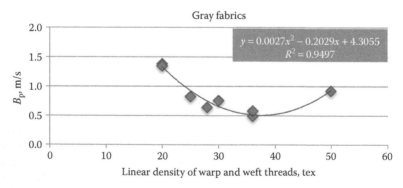

FIGURE 6.4

Influence of the yarns' linear density on the air permeability of gray fabrics.

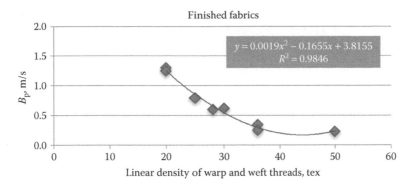

FIGURE 6.5
Influence of the yarns' linear density on the air permeability of finished fabrics.

of the mesostructure and air permeability coefficient B_p is of polynomial type and has a very good agreement with the experimental data ($R^2 = 0.95$).

After the desizing process, the parameters of meso- and macrostructure have been modified, and thus, the relation between Tt of the yarns and B_p has slightly changed, as can be seen in Figure 6.5. Sample 8 had a lower value of B_p in comparison with the gray fabric, as expected: the change in the warp and weft densities of the fabric, woven with ticker yarns, resulted in a reduction in the air permeability. The reason is that the reduced flow rate through the smaller pores of the macrostructure could hardly be compensated by the flow through the pores of the mesostructure, despite the small twists of the yarns with Tt = 50 tex.

The regression equation was again of second-order polynomial type, but it demonstrated even better correlation with the experimental data for the air permeability coefficient ($R^2 = 0.98$):

$$B_p = 0.0019Tt^2 - 0.1655Tt + 3.8155 \qquad (6.1)$$

where:
Tt is the linear density of warp and weft yarns

Certainly, Equation 6.1 is valid only for balanced in terms of linear density of the threads woven structures within the presented experiment.

6.4.2 Effect of the Twist

Figures 6.6 and 6.7 show the dependence between the twist of the mesostructure and the air permeability coefficient B_p for both gray and finished fabrics, respectively. The obtained results were in a good agreement with the expectation that the impact of the twist of warp and weft threads had to be analyzed together with the linear density, due to the presence of a technological link between the two characteristics of the mesostructure.

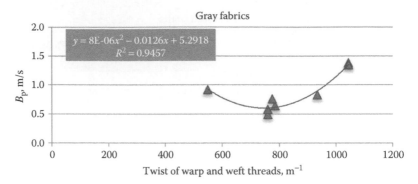

FIGURE 6.6
Influence of the yarns' twist on the air permeability of gray fabrics.

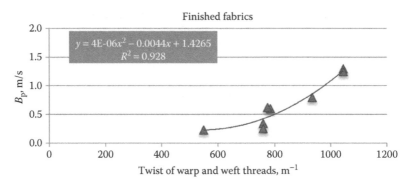

FIGURE 6.7
Influence of the yarns' twist on the air permeability of finished fabrics.

Theoretically, the increase in the yarns twist reduces the amount of air between the fibers in the mesostructure, that is, in the interfiber voids, through which a fluid flow can pass in through-thickness direction of the macrostructure. This leads to the logical expectation that the increase in the twist of the constituent yarns would reduce the air permeability of the woven macrostructure ceteris paribus.

The results in Figures 6.6 and 6.7 showed clearly, however, that the coefficient of air permeability B_p of the macrostructure rose with the twist of the yarns. The reason for this is the increased both compactness of the yarns with higher twist and pore areas between the yarns in the macrostructure. Highly twisted yarns are less hairy and their flow resistance is lower. The use of yarn sample 6 (Table 6.1) provoked again higher values of the air permeability coefficient. It can be assumed that there is a limit twist value before which yarns are sufficiently bulky so as larger flow rate to pass through the pores between the fibers in the mesostructure.

The experimental results in Figures 6.6 and 6.7 are described again with a second-order regression model. The correlation coefficient is a little higher for the gray samples ($R^2 = 0.95$) than for the finished ones ($R^2 = 0.93$), but both figures show a strong statistical relationship between studied characteristic of the mesostructure (twist of the yarns) and the air permeability coefficient. The regression equation (6.2) is again valid for a balanced fabric in terms of linear density within the performed experiment:

$$B_p = 4.10^{-6} T_y{}^2 - 0.0044 T_y + 1.4265 \tag{6.2}$$

In conclusion, it was found that the tested characteristics of the mesostructure, namely, yarns linear density and twist, affect the transmission of air in through-thickness direction of the studied woven macrostructures. The statistical relationships between them have very high correlation coefficients ($R^2 > 0.9$).

6.5 Analysis of the Effect of the Mesostructure on Heat Transfer through the Macrostructure

The results from the measurement of the conductive heat transfer through the woven macrostructures are summarized in Table 6.6: the thermal (conduction) resistance R_t and the thermal insulation I_{cl}. Only the finished fabrics were tested.

6.5.1 Effect of the Linear Density

The relationship between the linear density Tt of the mesostructure and the thermal insulation of the macrostructure is presented in Figure 6.8.

TABLE 6.6

Conductive Heat Transfer through the Woven Macrostructures

Sample	Thermal Resistance R_t, m²K/W	Thermal Insulation I_{cl}, clo
1	0.00399	0.026
2	0.004145	0.027
3	0.004145	0.027
4	0.00461	0.030
5	0.006005	0.039
6	0.009245	0.060
7	0.009795	0.063
8	0.009415	0.061

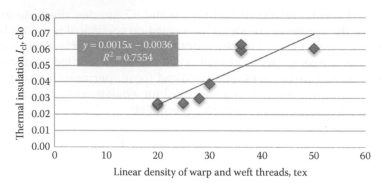

FIGURE 6.8
Influence of the yarns' linear density on the thermal insulation I_{cl} of the macrostructure.

The thermal insulation I_{cl} grew with the increase of Tt. The result is explained by the fact that the increase of Tt leads to an increase in thickness. It has been already discussed in Chapter 3 that the thickness of the textile macrostructure (regardless of its type—woven, knitted, or nonwoven) is considered to be the characteristic that preconditions mostly the heat transfer by conduction. This dependence was found to be directly proportional (Fanger 1972; Ukponmwan 1993; Wilbik-Halgas et al. 2006; Militky et al. 2010). At the same time, yarns of higher linear density comprise more air in the pores of the mesostructure, which increases the insulating ability of the macrostructure. The regression equation (6.3) for the relationship between yarns linear density and thermal insulation of the macrostructure has correlation coefficient $R^2 = 0.76$:

$$I_{cl} = 0.0015 Tt - 0.0036 \qquad (6.3)$$

6.5.2 Effect of the Twist

The graph on Figure 6.9 shows that the thermal insulation of the macrostructure decreased with the increase in the twist of the mesostructure. Explanations could be found in the increased pore area of the macrostructure and the amount of air between the fibers in the mesostructure that decreased with the higher twist values. Of course, an explanation could be the effect of the thickness of the macrostructure as well: yarns with more twists are thinner, and thus, the thickness of the macrostructure is smaller. The statistical relationship derived has a correlation coefficient $R^2 = 0.62$:

$$I_{cl} = -0.0004 T_y + 0.1074 \qquad (6.4)$$

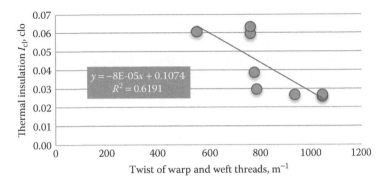

FIGURE 6.9
Influence of the yarns' twist on the thermal insulation I_{cl} of the macrostructure.

6.6 Analysis of the Effect of the Mesostructure on Moisture Transfer through the Macrostructure

The mean values from the testing of the finished woven samples are summarized in Table 6.7.

6.6.1 Effect of the Linear Density

In accordance with the WTT (wetting time top) characteristic, all samples had slow wettability as the wetting time was in the range 20–119 s (SDL Atlas 2010). At the same time, the penetration of fluid to the bottom surface in the transverse direction was faster, with the result that all the samples

TABLE 6.7

Moisture Transfer through the Woven Macrostructures

Sample	WTT, s	WTB, s	TSS, mm/s	BSS, mm/s	TAR,%/s	BAR,%/s	R_{index}, mm²/s	OMMC,-
1	30.64	7.45	0.31	4.9	89.23	6.32	2077.2	0.603
2	27.33	6	0.43	7.63	52.11	4.92	2498.9	0.700
3	82.06	1.87	0.1	4.33	29.54	9.97	2853.26	0.736
4	100	64.29	0	5.1	0	13.93	1116.3	0.581
5	81.48	2.98	0.13	3.13	11.26	3.66	2986.8	0.672
6	82.4	2.06	0.1	8.87	15.12	57.15	2942.9	0.764
7	64.84	3.21	0.18	2.9	18.98	4.64	2939.9	0.659
8	100	11.34	0	1.87	0	12.19	2024.45	0.590

(except sample 4) were defined as intermediate (samples 1, 2, and 8 with WTB = 5–19 s), fast (model 7 with WTB = 3–5 s), and very fast wettable (samples 3, 5, and 6 with WTB < 3 s).

The TSS values (top spreading speed) showed that all samples had very low speed of wetting the top surface, while the BSS values showed that the bottom surface wetted fast (BSS = 3–4 mm/s) to very fast (BSS > 4 mm/s). The only exception was sample 8, where the speed of wetting the bottom surface was assessed as slow (BSS = 1–2 mm/s).

Six of the macrostructures had very slow to slow absorption ratio of the top surface (TAR < 30%/s) and only samples 1 and 2 absorbed the liquid fast (TAR = 50–100%/s). Unlike the high speed of wetting of the bottom surface, almost all macrostructures had very low bottom absorption ratio (BAR < 30%/s). Sample 6 with high absorption ratio was the only exception.

The accumulative one-way transport capacity R_{index} reflects the transport of moisture in direction transversal of the sample thickness. It is calculated in accordance with the formula

$$R_{index} = \frac{A_b - A_t}{t}, \text{mm}^2/\text{s} \tag{6.5}$$

where:
A_b is the wetting area of the bottom surface, mm²
A_t is the wetting area of the top surface, mm²
t is the time of the testing, s

Unlike R_{index}, overall moisture management capacity (OMMC) reflects the moisture transfer in both transversal (by R_{index}) and lateral direction of the sample. OMMC varies in the range (0, 1) and an exemplary graph of the dependence of OMMC on the linear density of the mesostructure is shown in Figure 6.10.

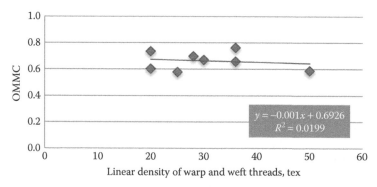

FIGURE 6.10
Influence of the yarns' linear density on the overall moisture management capacity.

TABLE 6.8

Summary of the Correlation Coefficients of the Linear Regression Equations for the Relationship between Yarn Twist and Linear Density and the Moisture Transport Characteristics

							Correlation Coefficient R^2	
			TSS,	BSS,			R_{index}	
	WTT, s	WTB, s	mm/s	mm/s	TAR,%/s	BAR,%/s	mm²/s	OMMC,-
Linear density	0.18	0.02	0.15	0.09	0.33	0.06	0.16	0.02
Twist	0.10	0.03	0.05	0.04	0.31	0.03	0.02	0.01

By analogy with the air permeability and heat transfer, the effects of the linear density and twist of the yarns on the transmission of moisture were sought. However, the charts and the statistical correlations showed very low values of R^2. The correlation coefficients for all the derived linear equations are summarized in Table 6.8.

6.6.2 Effect of the Twist

The same approach was used for the twist of the mesostructure: the experimental values of the tested characteristics for moisture transfer through the textile macrostructure were presented as a function of the yarns twist. The distribution of the single results was such that in most of the cases, the derived regression equations (of linear type) had very low correlation coefficient R^2 (Table 6.8). The relationship between yarns twist and OMMC is shown once again as a sample (Figure 6.11).

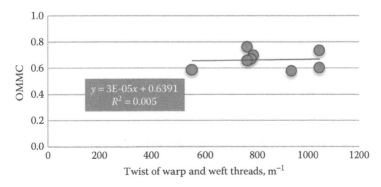

FIGURE 6.11
Influence of the yarns' twist on the overall moisture management capacity.

6.7 Summary

It was defined and proved the impossibility to produce identical woven macrostructures (i.e., with the same geometric, structural, and mass parameters), while changing the characteristics of the mesostructure—linear density and twist.

A systematic experiment was conducted to establish and assess the effect of the twist and linear density of the mesostructure on the transfer of air, heat, and moisture through the macrostructure.

The relationship between yarns linear density and twist on one side and air permeability of the macrostructures on the other was statistically proven and regression equations were derived with correlation coefficient $R^2 > 0.9$.

The relationship between yarns linear density and twist on one side and the thermal insulation of the macrostructures on the other was statistically proven and regression equations with high level of coincidence were derived.

The relationship between yarns linear density and twist on one side and the characteristics for moisture transport through the macrostructures on the other was not statistically proven through the regression analysis.

7

Experimental Investigation of the Macrostructure of Textiles for Clothing and Bedding

7.1 Introduction

The main aim of the study presented in this chapter was to perform a systematic experimental investigation on the effect of the characteristics of woven macrostructures for clothing and bed linen on their functional properties in the aspect of thermophysiological comfort. Gray and finished macrostructures were studied and the main characteristics of their meso- and macrostructures were determined experimentally.

Special attention was paid to the assessment of the porosity of the investigated samples and microscopic study of the pores of their macrostructure. A methodology for nondestructive analysis of the porosity of the fabric macrostructure was used.

The air permeability, heat transfer, and moisture transfer in through-thickness direction of the woven macrostructures were studied. The goal was to evaluate (to confirm or reject) the existence of a statistically proven relationships between structural, geometric, and mass properties of the macrostructures and the characteristics that describe the heat and mass transfer processes through them.

7.2 Properties of the Macrostructures

The cotton ring yarns, described in Chapter 6, were used for weaving of 14 samples on a Sulzer projectile weaving machine. The basic technological parameters of the gray fabrics—fabric weight, warp and weft densities, linear density of the yarns, weave and width of the fabrics—are shown in Table 7.1. Nine of the samples were woven in a plain weave, and the rest in a twill weave. Data in Table 7.1 were arranged in ascending order of the fabric weight of the samples.

TABLE 7.1

Basic Technological Parameters of the Gray Fabrics

Sample	Fabric Weight, g/m²	Warp Density, threads/ dm	Weft Density, threads/ dm	Tt_{wa}, tex	Tt_{wf}, tex	Weave	Width, mm	Thickness δ, mm
1	62	136	148	20	20	Plain	98	0.40
2	84	168	100	28	28	Plain	114	0.42
3	117	264	194	20	25	Plain	108	0.36
4	123	272	268	20	20	Plain	96	0.36
5	138	264	238	25	25	Plain	159	0.40
6	153	223	230	30	30	Plain	165	0.42
7	186	352	304	28	28	Twill 2/1	168	0.60
8	187	233	234	36	36	Plain	162	0.44
9	195	237	243	36	36	Plain	168	0.44
10	200	368	200	30	36	Twill 3/1	162	0.62
11	202	318	220	40	25	Twill 2/2	108	0.52
12	211	383	210	30	40	Twill 3/1	162	0.63
13	214	357	189	36	36	Twill 2/1	162	0.60
14	233	258	150	50	50	Plain	230	0.61

With a view to assess the porosity of the fabric, other structural character-istics of the gray fabrics were determined. Table 7.2 summarizes the experi-mental data from the measurement of warp and weft crimp, warp and weft cover factor, fabric cover factor E_s, and fabric areal porosity V_s.

The gray fabrics were treated for desizing in a bath of 25% solution of NaON (3 g/l) and 20% solution of textile auxiliaries (1 g/l) at 90°C for 1 h.

Tables 7.3 and 7.4 summarize the data for the finished fabrics, by analogy with Tables 7.1 and 7.2 for the gray fabrics.

Finishing caused expected changes in the parameters of the macrostruc-tures. Table 7.5 presents the percentage of increase/decrease in changes from the characteristic of the gray samples. The values of all characteristics, except the fabric weight, increased as a result of both desizing and relaxation of warp and weft threads.

7.3 Experimental Assessment of the Porosity

As noted in Section I, thermophysiological comfort related to textiles and clothing is directly dependent on the transfer of heat and mass through a single textile layer (macrostructure) or clothing (systems of textile lay-ers). In turn, properties like air permeability, moisture permeability, and

TABLE 7.2

Structural Parameters of the Gray Fabrics

Sample	Yarn Crimp, %		Linear Cover Factor E, %		Fabric Cover Factor E_s, %	Fabric Areal Porosity V_s, %
	Warp	Weft	Warp	Weft		
1	2.00	9.42	24.37	25.44	43.61	56.39
2	2.39	8.97	34.95	20.64	48.38	51.62
3	5.17	8.98	48.67	37.70	68.02	31.98
4	4.26	12.82	46.69	45.47	70.93	29.07
5	7.36	7.23	52.90	45.90	74.52	25.48
6	7.52	10.19	48.95	49.18	74.06	25.94
7	7.32	7.40	71.46	63.17	89.94	10.51
8	8.05	10.35	54.49	53.50	78.84	21.16
9	10.07	9.87	55.96	58.54	81.74	18.26
10	9.58	6.72	81.03	46.71	89.89	10.11
11	6.32	8.30	80.22	42.64	88.65	11.35
12	10.52	4.99	81.59	51.37	91.05	8.95
13	11.78	4.53	84.60	43.59	91.31	8.69
14	14.89	4.30	73.04	41.41	84.20	15.80

TABLE 7.3

Basic Technological Parameters of the Finished Fabrics

Sample	Fabric Weight, g/m²	P_{wa}, threads/ dm	P_{wf}, threads/ dm	Thickness δ, mm
1	63	142	166	0.43
2	89	176	124	0.45
3	118	272	218	0.39
4	133	284	294	0.39
5	138	270	266	0.41
6	157	234	254	0.47
7	184	383	338	0.68
8	201	258	238	0.49
9	197	252	260	0.48
10	202	386	226	0.71
11	206	324	254	0.64
12	212	402	240	0.71
13	219	376	216	0.65
14	235	267	166	0.62

TABLE 7.4

Structural Parameters of the Finished Fabrics

Sample	Yarn Crimp, %		Linear Cover Factor E, %		Fabric Cover Factor E_s, %	Fabric Areal Porosity V_s, %
	Warp	Weft	Warp	Weft		
1	11.39	10.75	24.41	27.57	45.26	54.74
2	15.54	10.19	36.02	25.06	52.04	47.96
3	12.89	9.34	48.75	41.00	69.76	30.24
4	10.83	13.31	48.88	53.12	76.03	23.96
5	13.19	12.01	52.70	49.38	76.06	23.94
6	14.24	13.27	49.51	53.83	76.69	23.31
7	15.07	11.93	59.07	68.11	86.95	13.05
8	16.42	13.19	63.31	52.81	82.69	17.31
9	15.54	11.62	57.66	61.68	83.78	16.22
10	18.96	9.79	81.95	50.79	91.12	8.88
11	15.36	9.54	78.25	50.89	89.72	10.28
12	11.97	9.09	79.25	57.54	92.00	8.00
13	21.20	8.21	85.59	49.85	92.77	7.23
14	23.75	7.27	72.27	45.09	84.77	15.23

TABLE 7.5

Percentage of Change of the Fabrics Characteristics after Finishing

Sample	Change, %					
	Fabric Weight	P_{wa}	P_{wf}	Thickness	Fabric Cover Factor	Fabric Areal Porosity
1	1.61	4.41	12.16	7.50	3.76	−2.93
2	5.95	4.76	24.00	7.14	7.59	−7.09
3	0.85	3.03	12.37	8.33	2.56	−5.44
4	8.13	4.41	9.70	8.33	7.19	−17.58
5	0.72	2.27	11.76	2.50	2.07	−6.04
6	2.61	4.93	10.43	11.90	3.55	−10.14
7	0.54	8.81	11.18	13.33	2.37	−13.35
8	7.49	10.73	1.71	11.36	4.88	−18.19
9	1.03	6.33	7.00	9.09	2.50	−11.17
10	1.00	4.89	13.00	14.52	1.37	−12.17
11	1.98	1.89	15.45	23.08	1.22	−5.90
12	0.47	4.96	14.29	12.70	1.04	−10.61
13	2.34	5.32	14.29	8.33	1.60	−16.80
14	0.86	3.49	10.67	8.20	0.68	−3.61

thermal insulation depend on the porosity of the macrostructure (Kothari and Newton 1974; Daukantienė and Skarulskienė 2005; Wilbik-Halgas et al. 2006, etc.).

A number of authors have dealt with studies and assessment of woven textiles porosity, both theoretical (Love 1954; Kemp 1958; Hamilton 1964; Peirce and Womersley 1978; Seyam and El-Shiekh 1993) and experimental studies (Dubrovski 2000; Dubrovski and Brezocnik 2002; Jakšic and Jakšic 2007; Zupin et al. 2012). The main reason for such studies is that the structure of woven fabrics, unlike that of the knitted and nonwoven textiles, is the most accurate and arranged like *tubular porous structure* (Elnashar 2005). Dubrovski (2000) developed a three-factor analytical model to determine the porosity of woven fabrics, and later, he (Dubrovski 2001) presented a mathematical model of porosity, based on the basic structural characteristics: linear density of the yarns, relative density of the fabric, and level of yarns interlacing. Militky et al. (2010) used the bulk density of the macrostructure for prediction of the air permeability of 27 fabrics, woven in a plain weave at a constant weft densities and variable warp densities. Although most of the publications have treated the pores between the yarns as cylinders with constant cross section (Jakšić 1975; Kulichenko and Langenhove 1992; Elnashar 2005; Xu and Wang 2005), the size of the pores between warp and weft threads, as well as their shape, is uneven. The same is valid for the distribution of the pores within the macrostructure (Cay et al. 2004). Woven macrostructures from the group of the so-called precise woven structures for medical purposes are an exception (Sefar 2008).

The most commonly used method for determination of the porosity of the woven macrostructures is by applying geometric relationships between the characteristics of the meso- and macrostructure: linear density (or diameter) of warp and weft threads, together with warp and weft densities. Thus, the processes of mass transfer, for example, can be coupled theoretically and experimentally with the fabrics porosity by using parameters such as average pore diameter and pore number per unit area (Lee et al. 2000; Cay et al. 2004).

Zupin et al. (2012) have concluded that pores are *the ideal parameter* for description of the porosity. In their research, they define four types of pores in the surface of single-layer woven macrostructures, depending on the four possible ways of interlacing between two warp and two weft threads. In fact, this is the model of Backer (1951) (Figure 7.1). Zupin et al. (2012) commented that in studies on permeability, the size of the pore must be equalized to the so-called hydraulic diameter and then pores of different shape could be represented as a pore with a circular cross section with the respective hydraulic diameter. Based on the experimental study of 36 woven samples with a total of 9 weaves, the authors found three characteristics of the woven macrostructure that showed a statistical link with a high degree of correlation with the air permeability: the hydraulic diameter of the pore, number of pores, and porosity. These findings are consistent with the results, presented in this chapter and in former publications (Angelova 2009b; Angelova 2012b).

FIGURE 7.1
Types of independent pores with interlacing threads. (Based on Backer, S., *Text. Res. J.*, 21, 703–714, 1951.)

Xu and Wang (2005) and Ogulata (2006) proposed a formula for calculating the area of the pore, used in the present study as well:

$$S_p = \left(\frac{100}{P_{wa}} - d_{wa} \right)\left(\frac{100}{P_{wf}} - d_{wf} \right) \tag{7.1}$$

where:
d_{wa} is the mean diameter of warp threads, mm
d_{wf} is mean diameter of weft threads, mm
P_{wa} is the warp density, threads/dm
P_{wf} is the weft density, threads/dm

A methodology, based on a nondestructive analysis, for the determination of the size of the pores in a woven macrostructure was developed and applied. A digital microscope Optika DM-15 with a built-in digital camera and software for automatic measurement of the area of closed contours was used. Twenty pictures were taken for each sample or a total of 280 images.
The methodology includes the following steps (Angelova 2012b):

- *Sample snapshots*: A digital microscope under 4× enlargement was used. Every single image included a different number of pores as the uneven distribution of warp and weft threads (most often due to the reed effect) preconditioned a different number of pores to be included in the visible part of the lens. The different densities of individual samples also led to differences in the number of pores in an image. Figure 7.2 illustrates the differences in the digital images, that is, for sample 4 (Figure 7.2a) and sample 14 (Figure 7.2b).

- *Transformation into a negative image*: As a result of preliminary experiments, it was found that the negative is more suitable for measuring the pore size, since the contrast of the filaments is greater than in the positive image. Figure 7.3 is a picture of sample 13 and the measurement of the pore size on the negative image.

- *Measurement of the pore area*: The voids between warp and weft threads are particularly irregular in size and shape. The experience within

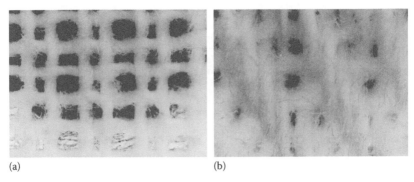

(a)   (b)

FIGURE 7.2
Microscopic images of (a) sample 4 and (b) sample 12 (4× enlargement).

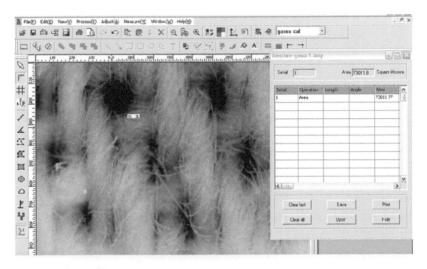

FIGURE 7.3
Measurement of a pore area, sample 13.

the experiment had shown that the operator tended to measure only the most visible pores, which could lead to large variations in the results. Therefore, a requirement was introduced to measure all the pores of the particular image. The measurement continued until reaching 100 values for pore area for each sample.

Table 7.6 presents the results from the measurements of the pore area, following the described methodology. The relative standard error is very high, due to the great uniformity of the pores in the macrostructures in size and area. The increase in the number of single measurements would undoubtedly reduce the relative standard error, but would increase both the

TABLE 7.6

Measured Area of the Pores in the Woven Macrostructures

Sample	Mean Pore Area, mm²	Variation Coefficient, %	Relative Standard Error, %
1	0.24282	198.6	25.8
2	0.25511	72.7	21.6
3	0.03051	173.6	22.5
4	0.03961	147.6	19.1
5	0.02411	130	16.9
6	0.04562	173	22.4
7	0.06250	72.7	21.5
8	0.02916	262	34
9	0.03783	176.8	22.9
10	0.02429	133.2	17.3
11	0.01423	137.1	17.8
12	0.03122	133.8	17.3
13	0.02259	153.2	19.9
14	0.02427	133.9	17.4

time and costs of the measurements. In all cases, the number of the measurements must be linked with the ultimate goal of the study and to evaluate whether additional costs are needed to obtain greater accuracy. This study showed, however, that the dispersion of the individual results in some samples was so big that its decrease would require additional hundreds of single measurements.

Figures 7.4 and 7.5 show exemplary results from the single pore area measurements of samples 2 and 13. The charts present the mean value and

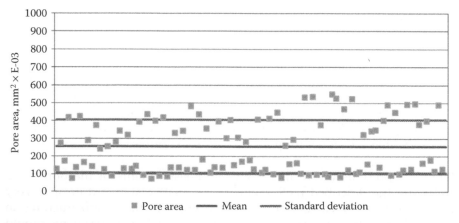

FIGURE 7.4

Sample 2—measurement of the pore area: single values, mean value, and standard deviation.

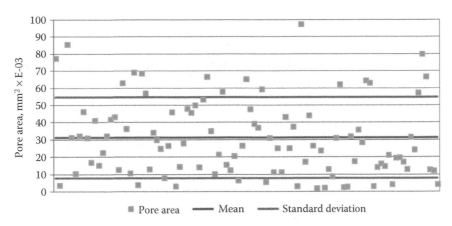

FIGURE 7.5
Sample 13—measurement of the pore area: single values, mean value, and standard deviation.

standard deviation for each sample. The results clearly show that the single values of the measured pore area are unevenly distributed and spread far beyond the limits of the standard deviation. The reason can be found in the non-uniform cross-sectional area of the fibers, the influence of the beat-up mechanism on the pore size, the instability of the woven structure, the hairiness of the yarns, and so on.

The shape of the pores had to be also determined so as to use the results from the pore area measurements for the creation of a virtual model of the woven macrostructure in Section III of the study. In this case, the pore could be equalized to a parallelepiped with a square cross-section or a cylinder (Angelova 2010c, 2010d). The equivalent side a_{eqv} or equivalent diameter d_{eqv} of the pore could be calculated from the following expressions:

$$a_{eqv} = \sqrt{S_{av}} \tag{7.2}$$

$$d_{eqv} = \sqrt{\frac{4S_{av}}{\pi}} \tag{7.3}$$

where:
S_{av} is the mean value of the pore area, mm^2

Data from experimental measurements were compared with the theoretical calculations for the pore area after the respective transformations to a pore with a circular or square cross section. The diameters of warp and weft threads were calculated in accordance with

$$d = 0.0357 \sqrt{\frac{Tt}{\gamma_m}} \tag{7.4}$$

where:

$\gamma_m = 0.815 \text{ mg/mm}^3$ is the mean density of the yarn

The results from the pore shape transformation and their comparison with the theoretical results are shown in Figures 7.6 and 7.7.

The analysis showed that the results from the measurements and theoretical calculation of the pore area ranged from almost equal values to values with substantial differences. The relative error determined by Equation 5.10 was less than 5% for samples 4, 6, 7, and 9. The relative error for samples 1 and 2 was between 5% and 10%. For all other samples, including samples 10–14,

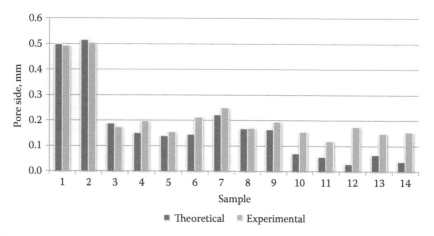

FIGURE 7.6
Square cross section of the pores: theoretical and experimental results for the pore side.

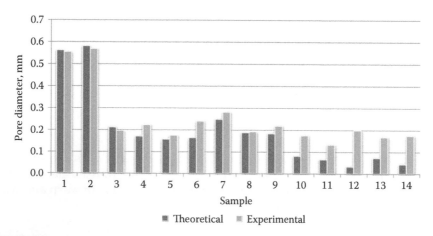

FIGURE 7.7
Circle cross section of the pores: theoretical and experimental results for the pore diameter.

the relative error was between 15% and 36%. A possible explanation may be found in different weave of the macrostructures: samples 1–9 were in a plain weave and the rest in twill.

The free floating of a thread over two or more threads from the other system in twill could result in more inaccurate positioning of warp and weft yarns in the woven macrostructure in comparison with the plain weave. Overall, the pores between the threads in the samples in twill were smaller. In addition, even under a microscope, pores were less visible, making the measurement more difficult. Xiao (2012) made similar conclusion, noting that there were even overlapping of the adjacent threads in tightly woven macrostructures. Another important conclusion of the author was that the pores in the relatively ordered woven macrostructure were of different size, but the knowledge on their geometry was very important for the evaluation of the transfer processes through the textile tissue.

A possible reason for the deviation could also be the theoretical determination of the diameter of the yarns, needed for the geometrical calculation of the pore area. One way of overcoming the possible incorrect value of the average density of the thread is an experimental measurement of the cross section and determination of the area of the yarn, occupied by fibers, compared to the total cross-sectional area of the yarns. Such an approach was used, for example, in the development of the analytical model of Gebart (1992), as well as in the studies of Saldaeva (2010) and Xiao (2012). The question of the effectiveness of the experimental study compared to the analytical calculation arises again, however.

Regarding the cross-sectional shape of the pore—round or square—the impact of the transformation was assessed by computational fluid dynamics (CFD) simulation, as presented in Section III of the book.

7.4 Experimental Assessment of the Air Permeability

Several published studies were dealing with the problem of the air permeability of textiles and clothing. Some of them were directly related to the evaluation of porosity and its impact on air permeability.

The investigations on air permeability of textiles dated back from the nineteenth century. Attempts to predict the transfer of air through different macrostructures began with the application of the law of Darcy, and later with the analytical relationship, known as the of Kozeny–Carman equation. The analytical models underwent intensive development in the twentieth century and several statistical models also appeared in parallel (Epps 1996).

The laws of Hagen–Poiseuille and Darcy were used in the work of Kulichenko and Langenhove (1992) to derive an equation that provided the theoretical link between the air permeability and structure of the fabric.

Darcy's law, however, is valid only for slow fluid flows, where the Reynolds number Re < 1. The analytical models based on it, are typically used to predict capillarity action or flows in micropores. The attempts to analytically calculate the air permeability, using Darcy's law, cannot be verified with results from the standard air permeability test for woven macrostructures, because the requirement for a pressure difference on both sides of the fabric in any event leads to transfer of a turbulent flow through the specimen.

The law of Hagen–Poiseuille was applied by Xu and Wang (2005), but the results were limited to the case of fabrics woven with monofilament. A theoretical calculation of the air permeability was reported in the work of Ogulata (2006). Hagen–Poiseuille's law is used in Section III of the book for modeling the air permeability of woven macrostructures from staple fiber yarns.

A detailed analysis of existing analytical models for calculation of through-thickness air permeability of woven macrostructures was presented in the studies of Saldaeva (2010) and Xiao (2012). Both authors made a compilation of models to predict the air permeability of the tested woven samples. The error between experimental and analytical results was very high, however. Xiao (2012) even used computer simulation as a tool for verification of the analytical calculation of the air permeability of airbags, provided that the numerical simulation itself needed verification.

Besides analytical models, other modern tools , such as neural networks (Tokarska 2004) and computer modeling (Nazarboland et al. 2007; Grouve et al. 2008), were used for predicting the air permeability. Fatahi and Alamdar (2010) examined the relationship between air permeability and mechanical properties of woven fabrics in plain weave. Special attention to the research on computer modeling of the transport of fluids through woven macrostructure is paid in Section III of the book.

The basic aim of a big part of the published experimental studies on air permeability was related to the study of porosity and description of the relationship between porosity and structural parameters (Goodings 1964; Kothari and Newton 1974; Daukantienė and Skarulskienė 2005; Wilbik-Halgas et al. 2006). The work of Elnasar (2005) and Militky et al. (2010) contributed to the knowledge in this field, together with the studies on the porosity and air permeability of woven macrostructures with high density (Rief et al. 2011).

Despite the importance of the topic, however, the relationship between specific geometrical, mass, and/or structural characteristics of fabrics and their air permeability is relatively rarely investigated. Perhaps the main reason is the problem of production of identical woven macrostructures, analyzed in Chapter 6. Many publications discussed methods and devices for measurement or simulation, as well as the relationship between air permeability and insulation abilities (Kulichenko and Langenhove 1992; Frydrych et al. 2002; Ogulata 2006; Simova et al. 2009; Fatahi and Alamdar 2010; Havlova 2010; Nabovati et al. 2010; Zupin et al. 2012). In terms of the characteristics of the macrostructure, however, it is assumed that thickness is the most important characteristic (Frydrych et al. 2002; Ogulata 2006; Zupin et al. 2012) and other characteristics are considered only as far as they are needed for proper description of the woven textile from technical or techno-logical point of view.

The main objective of this part of the study was to look for a characteris-tic of the woven structure that could be associated with its air permeabil-ity, using the methods of the regression analysis. A hypothesis was defined that air permeability would show a good correlation with a complex mass characteristic of the woven structures such as the fabric weight, as it reflects parameters of both meso- and macrostructure (warp and weft densities, fabric cover factor or areal porosity, thickness).

The air permeability of the woven macrostructures was measured in accordance with the method described in Chapter 5. Both gray and finished fabrics were tested so as to assess the effect of the changed characteristics of the macrostructure (see Table 7.6) on the transport of fluids in through-thickness direction of the samples. The results from the air permeability measurements are shown in Table 7.7: the flow rate through a sample of 10 cm^2 and the calculated air permeability coefficient B_p.

The graphs in Figures 7.8 through 7.13 refer to the air permeability coef-ficient of the gray fabrics. The graphs in Figures 7.14 through 7.19 are similar, but show the results from the measurements of the finished samples.

TABLE 7.7

Results from the Air Permeability Measurements

Sample	Air Flow Rate, m^3/s		Air Permeability Coefficient B_p, m/s	
	Gray Fabrics	Finished Fabrics	Gray Fabrics	Finished Fabrics
1	0.00276	0.0026	1.38	1.30
2	0.0027	0.0025	1.35	1.25
3	0.00268	0.00264	1.34	1.32
4	0.00166	0.00158	0.83	0.79
5	0.00128	0.0012	0.64	0.60
6	0.00152	0.00124	0.76	0.62
7	0.00196	0.00062	0.98	0.31
8	0.00098	0.00068	0.49	0.34
9	0.00116	0.0005	0.58	0.25
10	0.00188	0.00076	0.94	0.38
11	0.00168	0.00052	0.84	0.26
12	0.00168	0.00058	0.84	0.29
13	0.00144	0.00056	0.72	0.28
14	0.00184	0.00046	0.92	0.23

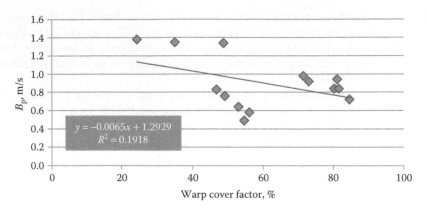

FIGURE 7.8
Air permeability as a function of the warp cover factor of the macrostructure, gray fabrics.

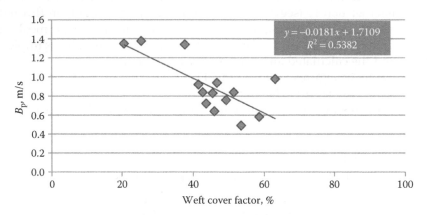

FIGURE 7.9
Air permeability as a function of the weft cover factor of the macrostructure, gray fabrics.

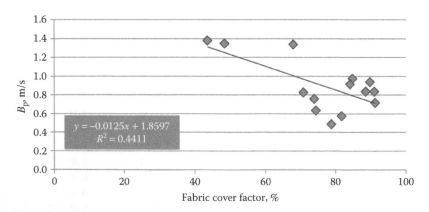

FIGURE 7.10
Air permeability as a function of the fabric cover factor of the macrostructure, gray fabrics.

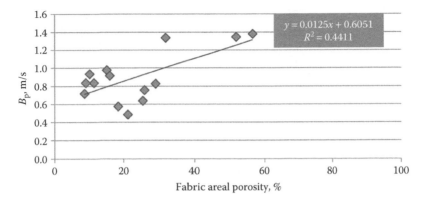

FIGURE 7.11
Air permeability as a function of the areal porosity of the macrostructure, gray fabrics.

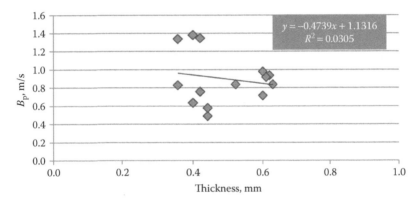

FIGURE 7.12
Air permeability as a function of the thickness of the macrostructure, gray fabrics.

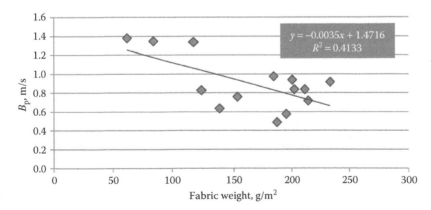

FIGURE 7.13
Air permeability as a function of the fabric weight of the macrostructure, gray fabrics.

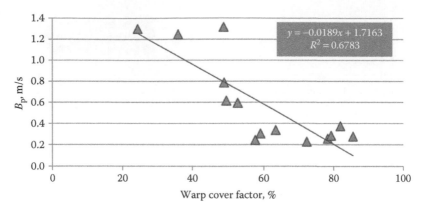

FIGURE 7.14
Air permeability as a function of the warp cover factor of the macrostructure, finished fabrics.

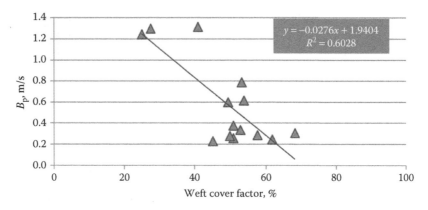

FIGURE 7.15
Air permeability as a function of the weft cover factor of the macrostructure, finished fabrics.

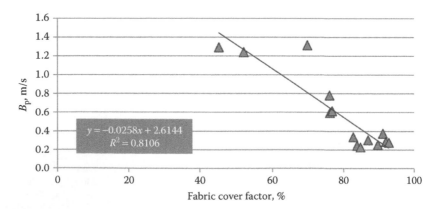

FIGURE 7.16
Air permeability as a function of the fabric cover factor of the macrostructure, finished fabrics.

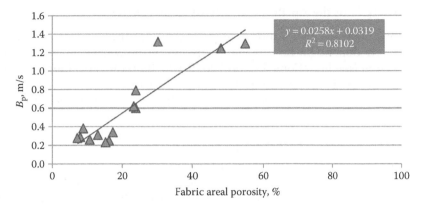

FIGURE 7.17
Air permeability as a function of the areal porosity of the macrostructure, finished fabrics.

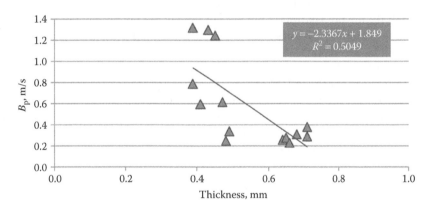

FIGURE 7.18
Air permeability as a function of the thickness of the macrostructure, finished fabrics.

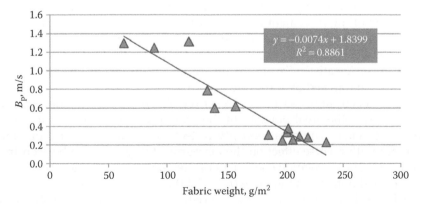

FIGURE 7.19
Air permeability as a function of the fabric weight of the macrostructure, finished fabrics.

7.4.1 Analysis of the Air Permeability of Gray Fabrics

The fabric cover factor E_s and areal porosity V_s of the macrostructure are functionally dependent on the determination of the warp E_{wa} and weft E_{wf} cover factors (see Equations 5.7 and 5.8). It was therefore considered appropriate to assess the impact of each of these structural characteristics separately. Figure 7.8 shows the change of the coefficient of air permeability B_p as a function of the warp cover factor E_{wa}. An analogous dependence is shown in Figure 7.9 for the weft cover factor E_{wf}.

It is logical to expect that the linear (warp or weft) cover factor, which evaluates the simultaneous effect of the threads density in the respective direction and yarn linear density, is a more appropriate structural characteristic to be used instead of warp or weft density, defined in the literature as especially important. Linear cover factor actually reflects the joint influence of characteristics of the macrostructure (number of threads in a certain distance) and mesostructure (linear density or thickness of the constituent threads). It can be assumed that the increase in the linear cover factor in one of the system of threads (warp or weft) will lead to decrement of the air permeability coefficient.

The results in Figures 7.8 and 7.9 showed the expected trend. However, the correlation coefficient of the regression equation for the relationship between B_p and the warp cover factor was very low ($R^2 = 0.19$). The correlation coefficient for the weft cover factor's influence was higher ($R^2 = 0.54$). The additional statistical proof by the linear correlation coefficient showed that there was a statistically proven relationship between the air permeability coefficient B_p and warp cover factor ($t_R = 3.94 > t_T = 2.56$). The statistical relationship between B_p and the weft cover factor was also proven ($t_R = 7.83 > t_T = 2.56$).

Certainly, the fabric cover area is a more complex characteristic, assessing the impact of the two characteristics of the macrostructure (warp and weft densities) and the two characteristics of the mesostructure (linear density of warp and weft threads). The trend on Figure 7.10 was logical: B_p decreased with linear filling increment in both directions, that is, with the increment of the fabric cover factor. The regression equation had a low coefficient of correlation ($R^2 = 0.44$), due to the low value of R^2 for the regression equation in Figure 7.8. A statistically proven relationship between B_p and fabric cover factor was guaranteed by the student's criteria.

Similar discussion has to be presented for the effect of the opposite characteristic—the areal porosity of the macrostructure—on the air permeability coefficient. The areal porosity reflects the same structural phenomenon, but with reverse effect on the air permeability, as it can be seen from Figure 7.11. The correlation coefficient of the regression equation between B_p and the areal porosity is the same ($R^2 = 0.44$) as in Figure 7.10.

Thickness is an important characteristic of the textile macrostructures (woven, knitted and nonwoven), which is considered determinative for the thermal insulation properties of textiles and clothing. This relationship was discussed in Section I and in Chapter 6, as well as is underlying in the annexes of the ISO 9920 (1995) international standard. The ability of any macrostructure to trap a layer of air between the human body and the innermost layer of clothing, however, is associated primarily with the property of *breathability*. In this sense, it was expected a clear relationship between the thickness and air permeability coefficient of the woven macrostructure to be established in this part of the study. Figure 7.12 showed, however, that the linear relationship between the thickness of the samples and B_p had close to zero correlation coefficient, and the distribution of single results was too chaotic.

The fabric weight is implicitly a complex mass characteristic of the woven structures, reflecting both characteristics of meso- and the macrostructure. The trend in Figure 7.13 was logical: the air permeability coefficient B_p decreased with the increase in the samples weight. The correlation coefficient of the linear regression equation is low ($R^2 = 0.41$), but the relationship between the two variables is statistically proven by the student's coefficient ($t_R = 6.24 > t_T = 2.56$).

7.4.2 Analysis of the Air Permeability of Finished Fabrics

The graphs in Figures 7.14 and 7.15 show the influence of warp and weft cover factor, respectively, on the coefficient of air permeability of the finished fabric. The linear regression equations had higher correlation coefficients in comparison with the analogous dependences for the gray fabrics:

- For the warp cover factor:

$$B_p = -0.0189E_{wa} + 1.7163 \quad \left(R^2 = 0.68\right) \tag{7.5}$$

- For the weft cover factor:

$$B_p = -0.0276E_{wf} + 1.9404 \quad \left(R^2 = 0.60\right) \tag{7.6}$$

Figures 7.16 and 7.17 present the effect of the fabric cover factor and areal porosity on the air permeability. The established regression equation (7.7) for the relationship between B_p and the areal porosity had $R^2 = 0.81$, which was quite higher in comparison with the respective correlation coefficient for the gray fabrics ($R^2 = 0.44$):

$$B_p = 0.0258V_s + 0.0319 \tag{7.7}$$

The dependence between B_p and the thickness of the finished fabrics is shown in Figure 7.18. The correlation coefficient of the linear regression equation was higher ($R^2 = 0.5$) than in the case of the gray samples (Figure 7.12), but still remained low so as the statistical dependence to be considered strong enough.

The relationship between the fabric weight and the air permeability coefficient is shown in Figure 7.19. The high correlation coefficient ($R^2 = 0.89$) allowed to assess the resulting statistical dependence as very good. It proved the hypothesis that the air permeability of the woven macrostructure will show a good correlation with a complex mass characteristic like the fabric weight. The derived regression equation is

$$B_p = -0.0074m_s + 1.8399 \qquad (7.8)$$

Explanations of the better grouping of the single results for the finished fabrics, compared to the gray ones, can be found in the removal of the sizing agent from the warp threads and the overall relaxation of the internal stresses in the threads after the wet finishing process. The presence of a sizing agent on the warp threads changed the dynamic model of the air flow through the mesostructure of the gray samples. At the same time, the areal porosity of the gray fabrics was higher, which logically determined the largest air flow rate through a unit area.

7.5 Experimental Assessment of the Conductive Heat Transfer

Henry (1939) developed the first model that described the mechanisms of heat transfer through a hygroscopic textile barrier. The model was entirely analytical and provided a linear relationship between the moisture content of the textile material and temperature. The model, however, was not verified with experimental data.

The work of Henry (1939), together with the increased interest in thermal comfort, created the basis for sustainable experimental and theoretical research in the field, especially for the development of more accurate analytical models (Li and Holcombe 1992, 1998; Ding et al. 2010).

In the last decades, intensive research on the problem of transfer of heat through a textile layer and systems of layers, as well as through clothes for different purposes, including protective clothing, was carried out (Morse et al. 1973; Watt and Darcy 1979; Farnworth 1983; Williams and Curry 1992; Torvi 1997; Song et al. 2004; Ding et al. 2011; Li et al. 2013).

Morse et al. (1973) developed a theoretical model of the heat transfer based on an experimental study of macrostructures for protective clothing. Watt and Darcy (1979) performed a series of experiments to determine the sorption isotherms of wool in the range 20°C–100°C, evaluating the join effect of the transfer of moisture and heat through the textile barrier. Farnworth (1983) studied two mechanisms of heat transfer through a textile structure: conduction and radiation. The comparison between theoretical and experimental data showed significant deviations, which the author attributed to the different absorption capacity of the porous medium. Later on, he created a mathematical model for combined transport of water vapor and heat through a textile layer (Farnworth 1986), which (model) became a background for the development of other theoretical models (Saldaeva 2010; Ding et al. 2010).

Holcombe and Hoschke (1983) conducted a large-scale experimental study of knitted macrostructures for clothing, a total of 31 samples of different composition and characteristics, brought together by their common use. The authors measured the thermal resistance and established a strong influence of thickness, less impact on other structural parameters, and very little effect on the type of the fibers. In fact, the heat transfer depends directly on the thermal conductivity of the fibers and it seems reasonable to assume that the composition of the textiles would have the greatest significance for their thermal insulation properties. Behera and Hari (2010) also paid special attention in their monograph to this issue. They have discussed that most raw materials for the production of textiles have close values of thermal conductivity and the effect of the type of fibers on the thermal insulation is small. Another important reason is the far more significant impact of structural, geometric, and mass characteristics of the textile on the heat insulating capacity in comparison with the type of fibers.

Many studies have drawn this important conclusion, but large-scale studies like that of Holcombe and Hoschke (1983) on knitted macrostructures are rare, especially for woven macrostructures. Due to the significant difference in structural, geometric, and mass characteristics of textiles with different application, the authors combined the textiles in their studies taking into account most often the indicator *application*. Lis et al. (1962), for example, presented a comprehensive report on the thermal conductivity of fabrics for parachutes from polyamide. The resistance of the woven macrostructure to different loads in close to operational conditions was also investigated. Abbott (1973) and Baitinger (1979) dealt with the behavior of textiles in combustion tests and mechanisms to slow the accumulation of heat in textiles. Benisek et al. (1979) studied fabrics for protective clothing of different materials and found that the density and thickness of the textile macrostructures had a direct impact on the protection from convective heat sources. Torvi (1997) also studied the effect of structural characteristics on the protective properties of single layer fabrics for protective clothing. Xiao (2012) investigated the heat transfer through very tight woven macrostructures, including deformed ones, used for airbags.

Sun et al. (2000) showed that the thermal resistance of textiles depended mainly on the composition and structural characteristics and presented regression models to establish the effect of the fiber type, thickness of the fabric, and the number of layers in the system. The results were similar to those of Epps (1988), who tested experimentally the air permeability and thermal conductivity of 10 samples of woven, knitted, and nonwoven macrostructures. However, neither the combinations of different types of macrostructures (i.e., woven and nonwoven) were studied nor statistical dependencies (although it was possible) were derived.

Later, the same author continued his research in the work (already commented in Chapter 6) of Epps and Song (1992). They studied three macrostructures in a plain weave with a fabric weight of 131, 152, and 216 g/m^2. Regression equations for the effect of the number of the layers on the transfer of heat and air were proposed. The authors concluded, however, that within the performed experiment they didn't found geometric or structural characteristic of the woven samples, which affected the heat transfer or air permeability. The conclusion was logical, having in mind the small number of tested samples.

A similar conclusion, but with more samples, was reported earlier by Yoon and Buckley (1984). They investigated five woven macrostructures from cotton, polyester, and cotton/polyester blend in a different ratio with respect to the transfer of air and heat in a through-thickness direction of the fabrics. The authors ascertained that the porosity and thickness determined to the greatest extent the air permeability, transmission of water vapor and heat. They found out that the type of fibers is important for the transport of liquids only, where the cotton yarns had an advantage compared to polyester ones. At the same time, the authors ascertained that the geometric parameters influenced this process as well, so that the transfer of liquid through the CO/PES 50/50% samples was almost the same as in CO 100%.

A study of the transfer of moisture and heat from the body to the environment was presented in the work of Ding et al. (2010). The parameters of the environment were controlled and two types of textile layers were investigated: denim and Nomex®. The authors measured the thermal conductivity and developed an analytical model that was an improvement of the model of Min et al. (2007), as forced convection was also considered. The model was verified experimentally and showed good coincidence with the experimental results.

Another direction of the research on heat transfer processes has been developed in the last years, dedicated to studies of clothing as complex items. Thermal mannequins and field measurements with the participation of real subjects are used. This type of experiments go beyond the scope of the study in Section II, focused on woven textiles only, but it is worth mentioning the work of Li et al. (2013), who examined the thickness of the layer between the garment and the human body by 3D body scanner. The authors tested 35 shirts with different body fitting, made of seven woven macrostructures, and established regression models for the relationship between the insulating ability of the garment and the size of the air layer between the skin and clothing.

Despite the intensive research in the last decades, several authors of dissertations or monographs have drawn the conclusion that there is still lack of investigations in the field of heat transfer in through-thickness direction of textile macrostructures, especially thin ones (Behera and Hari 2010; Saldaeva 2010; Xiao 2012).

The assessment of the heat transfer through a textile layer is important for the conditions of standard indoor environment not because of the existence of a large or rapidly changing temperature gradient, but because of the need of the textiles to ensure the maintenance of the core body temperature in a very narrow range—around 37°C. The knowledge of thermal insulation is important also for numerical simulation of heat transfer from the body to the environment through a thermophysiological model, presented in Section IV. All 14 finished samples were investigated in accordance with the methodology described in Chapter 5.

The results from the conductive heat transfer measurements are presented in Table 7.8: the thermal resistance R_t and thermal insulation I_{cl} (clo). A regression analysis was performed similar to the analysis on the air permeability in order to establish a correlation between characteristic of the macrostructure and the thermal insulation I_{cl}. The thermal insulation was preferred to the thermal resistance since it was used in the simulations with a thermophysiological model in Section IV. Both variables, however, are functionally dependent.

The graph of Figure 7.20 depicted the I_{cl} values as a function of the weight m_s of the samples. The increase in the fabric weight caused an increase in the thermal insulation due to both the presumed increase in the thickness and the retention of a greater amount of air in the pores of the meso- and macrostructure. The linear relationship had a correlation coefficient $R^2 = 0.78$:

TABLE 7.8

Results from the Measurement of the Thermal Insulation of the Samples

Sample	Thermal Resistance R_t, m²K/W	Thermal Insulation I_{cl}, clo
1	0.004	0.026
2	0.004	0.027
3	0.004	0.024
4	0.004	0.027
5	0.005	0.030
6	0.006	0.039
7	0.011	0.070
8	0.009	0.060
9	0.010	0.063
10	0.011	0.068
11	0.011	0.071
12	0.011	0.068
13	0.009	0.059
14	0.009	0.061

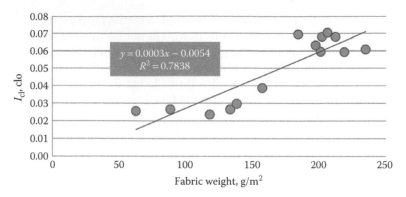

FIGURE 7.20
Thermal insulation as a function of the fabric weight of the macrostructure.

$$I_{cl} = 0.0003m_s - 0.0054 \qquad (7.9)$$

The effect of the thickness requires special assessment, as a number of authors (i.e., Morris 1955; Fanger 1972; Ukponmwan 1993) commented the presence of a linear relationship between the thickness of a single textile layer and its insulating properties. Figure 7.21 clearly shows the linear dependence established between the two characteristics of the investigated samples: the correlation coefficient $R^2 = 0.78$, which is in the range of R^2 for the statistical relation between I_{cl} and the fabric weight (Figure 7.20):

$$I_{cl} = 0.137\delta - 0.0242 \qquad (7.10)$$

The relationship between the thermal insulation I_{cl} and the areal porosity of the macrostructure is presented in Figure 7.22. The analysis of the graph showed that the I_{cl} values remained relatively constant for $V_s > 30\%$. An explanation could be that beyond a certain point the thermal insulation of a

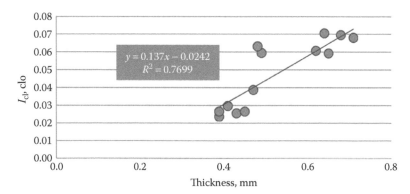

FIGURE 7.21
Thermal insulation as a function of the thickness of the macrostructure.

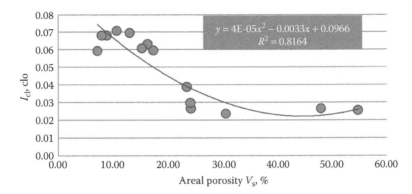

FIGURE 7.22
Thermal insulation as a function of the areal porosity of the macrostructure.

fabric with higher areal porosity is only due to the insulating ability of the fibers, and not to the air, retained between them or in the pores between the yarns. Unfortunately, the derived regression equation is not able to reflect the physical nature of the phenomenon, as it is a second-order polynomial and suggests the presence of extreme, but not a straight line for the points after $V_s > 30\%$. The correlation coefficient is high ($R^2 = 0.82$) and demonstrates clearly the correlation between the areal porosity and thermal insulation, which is not reported so far in the literature:

$$I_{cl} = 4.10^{-5}V_s^2 - 0.0033V_s + 0.0966 \tag{7.11}$$

Certainly, all regression equations are valid only for the performed experiments and the particular woven macrostructures.

7.6 Experimental Assessment of the Moisture Transport

All finished fabrics were tested following the procedures described in Chapter 5. Table 7.9 summarizes the mean values of the measurements for the eight individual characteristics of the moisture transfer: wetting time of the top (WTT, s) and bottom (WTB, s) side of the specimen; spreading speed of the top (TSS, mm/s) and bottom (BSS, mm/s) surface; top (TAR,%/s) and bottom (BAR,%/s) absorption ratio; accumulative one-way transport index (R_{index}, mm²/s); and the overall moisture management capacity (OMMC).

The analysis of the results for WTT showed that only samples 12 and 13 could be determined as moderately fast wetting (WTT < 20), and all others wet slowly. The results for WTB characteristic indicated that samples 3, 5, 7, and 10–13 were again assessed as slowly wetting and the others as fast and even very fast wetting (samples 4, 6, 8).

TABLE 7.9

Moisture Transport through the Finished Macrostructures

Sample	WTT, s	WTB, s	TAR, %/s	BAR, %/s	TSS, mm/s	BSS, mm/s	R_{index}, mm²/s	OMMC
1	30.64	7.45	89.23	6.32	0.31	4.9	2077.2	0.603
2	27.33	6	52.11	4.92	0.43	7.63	2498.9	0.700
3	100	60.08	0	27.51	0	11.09	998.12	0.665
4	82.06	1.87	29.54	9.97	0.1	4.33	2853.26	0.736
5	100	64.29	0	13.93	0	5.1	1116.3	0.581
6	81.48	2.98	11.26	3.66	0.13	3.13	2986.8	0.672
7	100	100	0	0	0	0	1342.5	0.500
8	82.4	2.06	15.12	57.15	0.1	8.87	2942.9	0.764
9	64.84	3.21	18.98	4.64	0.18	2.9	2939.9	0.659
10	100	100	0	0	0	0	735.8	0.485
11	100	46.88	0	2.55	0	0.03	1592.95	0.500
12	9.83	30.45	173.69	12.93	0.59	0	773.57	0.523
13	11.68	20.64	206.9	16.48	0.45	1.64	1231.23	0.618
14	100	11.34	0	12.19	0	1.87	2024.45	0.590

Similar analyzes can be made for each of the measured characteristics of the moisture transport. Samples 5, 7, 10, 11, and 14 could be assessed as macrostructures with very slow wetting or without wetting, taking into account the values for TAR, BAR, TSS, and BSS.

The accumulative one-way transport index, however, determined all studied samples as excellent in transferring moisture ($R_{index} > 400$). At the same time, the results for the OMMC, which is a general characteristic, allowed to assess samples 5, 7, 10–12, and 14 as macrostructures with good wetting, and all other samples as macrostructures with very good wetting. To some extent, this result could be associated with the higher fabric weight of samples 10–14.

The fabric weight, thickness, and fabric cover factor (or the areal porosity) are the main parameters of the woven macrostructures that can be expected to influence the moisture transport through the samples.

The graphical representation of the results from the experiment and the regression analysis performed showed, however, that the correlation coefficients R^2 of the derived linear regression equations were very low, as summarized in Table 7.10. The highest correlation coefficient was achieved for the linear relationship between the thickness and BSS. The proof with the linear correlation coefficient showed that the relationship between the areal porosity and BSS was statistically significant as well. The statistical relation between the thickness and OMMC had also high correlation coefficient ($R^2 = 0.56$).

Figures 7.23 and 7.24 present sample graphs for the effect of thickness of the macrostructures on BSS and OMMC, respectively. The derived regression equations and the respective correlation coefficients are shown in the figures.

TABLE 7.10

Summary of the Correlation Coefficients for the Linear Relationship between the Characteristics of the Macrostructure and the Characteristics for Moisture Transfer

			\multicolumn{6}{c}{Correlation Coefficient R^2}				
	WTT, s	WTB, s	TSS, mm/s	BSS, mm/s	TAR, %/s	BAR, %/s	$R_{index'}$ mm²/s	OMMC,-
Fabric weight	0.03	0.02	0.01	0.35	0.01	0.01	0.04	0.13
Thickness	0.01	0.20	0.02	0.64	0.09	0.07	0.28	0.56
Areal porosity	0.06	0.11	0.03	0.37	0.01	0.01	0.11	0.17

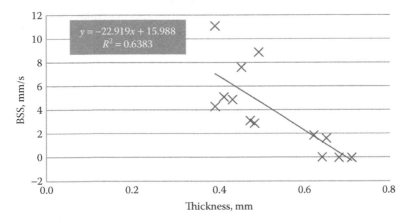

FIGURE 7.23
Influence of the thickness on the bottom spreading speed.

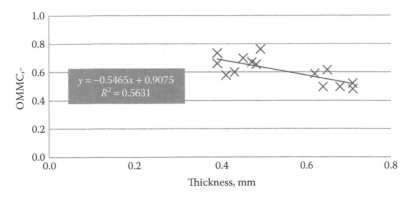

FIGURE 7.24
Influence of the thickness on the overall moisture management capacity.

7.7 Summary

A systematic, comprehensive experimental study on the effect of the characteristics of the macrostructure of fabric for clothing and bedding on the transfer of air, moisture, and heat was performed.

A methodology based on a nondestructive analysis of woven samples for the determination of the pore size of the macrostructures was developed and applied. It was found that the pores in the woven macrostructures differ substantially in shape and size within the same sample.

It was shown by means of correlation analysis that the fabric weight could be considered as a complex characteristic of the textiles, as it reflected other basic characteristics of both meso- and macrostructure (warp and weft densities, fabric cover factor or areal porosity, thickness).

A hypothesis that the air permeability had a good correlation with a complex characteristic of the woven macrostructure as the fabric weight was defined and proven.

It was shown within the experiment that the tested characteristics of the meso- and macrostructure, excluding weft density, affected the air permeability.

New facts on the existence of a statistical relationship between the air permeability coefficient B_p and characteristics of the macrostructure—areal porosity (fabric cover factor) and fabric weight—were determined. Regression models of linear type with correlation coefficients of 0.81 and 0.89, respectively, were ascertained.

A linear relationship between the thickness of the macrostructure and its thermal insulation I_{cl} was found, confirming results from other studies.

The existence of a relationship between the thermal insulation I_{cl} and the areal porosity (fabric cover factor) was ascertained; such a relationship has not been reported in the literature.

Lack of dependences between the studied characteristics of the moisture transport and the characteristics of the woven macrostructure with high enough correlation coefficient ($R^2 > 0.7$) was found.

It was also found that the fabric weight is a complex characteristic of the woven textiles that reflects simultaneously parameters of both mesostructure and macrostructure and preconditions the processes of transfer of air and heat through the macrostructure.

8

Experimental Investigation of the Macrostructure of Textiles for Surgical Clothes and Medical Drapes

8.1 Introduction

As medical products, textiles used in operating rooms and medical establishments must comply with the requirements of the European Medical Devices Directive 2007/47/EC. The conformity is dependent on the type of the medical product and, in particular, of the class to which it belongs. The European Medical Devices Directive has defined four classes of risk (the fourth one indicates the highest risk of infection).

The original function of surgical clothing was to protect the patients from the influence of the surgical team. Therefore, the only requirement to the surgical clothes was to be produced from relatively porous and permeable cotton fabrics with soft feel. Gradually, the role of surgical clothing changed with the study of the hazards, associated with blood-borne pathogens—to the already known need to protect the patient from the surgeon, the need to protect the hospital staff from the patient was added.

The textiles, used in the operating rooms, belong to class 1 (low risk) medical devices and must meet the various parts of the standard EN 13795 (2006). This standard is written in the spirit of the European Directive and aims to prevent the transmission of infections between medical staff and patients during surgery and other invasive interventions, as well as nosocomial infections.

EN 13795 (2006) is dedicated to the performance requirements for surgical gowns, surgical drapes, and clean air suits used as medical devices. In Parts 1 and 2 of the standard, all the requirements (and test methods) associated with the transmission of infection from patient to hospital staff and vice versa were set. The standard, however, does not deal with thermophysiological comfort either of the medical staff or of the patient. The interaction between the human body and the indoor environment of an operating room is different for the patient and the medical staff, and therefore, the adherence to ISO 7730 (1995) and ISO 9920 (1995) requirements does not necessarily provide the thermophysiological comfort of all *inhabitants*.

8.2 Thermophysiological Comfort in an Operating Room

The ISO 7730 (1995) International Standard does not make difference between the interior spaces in terms of thermal comfort; therefore, any modern operating room, for example, must meet the requirements of this standard, like an office space and theater. At the same time, the *inhabitants* of the operating room have different requirements for the environmental parameters, because of the different roles they have: surgeons, anesthesiologists and nurses, patients.

Woods et al. (1986) divided the surgical room into three zones:

- *Zone 1* (micro zone): This zone is directly under the operating lights; this is the place of surgeons and patients.

- *Zone 2* (sterile zone): This zone is where surgical instruments and apparatus are, immediately after zone 1. Surgical nurses move between zones 1 and 2.

- *Zone 3* (mini zone): This zone is the less sterile area; this is the workplace of nurses, anesthesiologists, and technicians, needed to work with medical equipment.

It has been found that in the area of the operating table, just below the lights, the air temperature is higher on average by 3°C than the rest of the room (Chow et al. 2006). At the same time, almost all parts of the body of the surgeons are covered by protective clothing (except the eyes). This practically means that the heat exchange between the surgeon's body and the environment is carried out only through a barrier of the textile layer/layers. The latter is coupled with relatively static activity, but performed under mental tension and stress.

The analysis shows that even when the air conditioning system provides the standard conditions for thermal comfort in the operating room, probably only the anesthesiologist and nurses will be able to define the thermal environment as comfortable. Therefore, special attention has to be paid to the thermophysiological comfort of the patient and the surgeons.

8.2.1 Surgeons

The materials used for the production of surgical clothes and drapes are different, depending on the type of their macrostructure: woven or nonwoven. Chemical fibers dominate, however, in both woven and nonwoven textiles: polyester, polypropylene, and membranes of polytetrafluoroethylene and polyurethane. Common combination is a mixture of cotton and polyester, with a majority participation of polyester fibers. Hohenstein Test Report (2000) informed that clothing with cotton/polyester composition received higher voting during field measurement by subjective evaluation, though, in terms of the risk of infection, this fiber blend is not particularly suitable for surgical clothing.

As a rule, the system of clothing of a surgeon in the operating room includes (Song 2011)

- Personal underwear and socks made of cotton fibers or blends of cotton and chemical fibers, non-sterile.
- Pants and short sleeves shirt (blouse), blend of cotton and polyester.
- Sterile surgical gown (mantle) with *standard* or *high* protection class (EN 13795, 2006).
- Mask, hat, gloves, and shoes.

The human thermophysiological comfort in the operating room, like in all other indoor spaces, depends on the factors listed in Chapter 1, with the following peculiarities:

- *Parameters of the environment (temperature, humidity, etc.)*: They provide the thermophysiological comfort of the surgeons to a relative degree as the environmental parameters around the operating table are strongly influenced by the radiation heat from the surgical lights.
- *Metabolism*: It is not amenable to control by the surgeons. They perform stationary activity done in upright or slightly bent position, which further leads to physical stress (except for some surgery in ophthalmology and neurosurgery, where surgeons can be seated). High levels of stress are added to the continued work in an awkward posture, which particularly affect the endocrine glands and the related thermal regulation system of the body (Nocker 2011).
- *Clothing*: They are most frequently multilayered with a complex effect on the body.

To this set of factors, Nocker (2011) added the individual status, premised on factors such as lack of sleep, nutrition, stimulating drinks, alcohol and drugs, blood sugar levels, physical fitness, and motivation. Randall (1946) analyzed the reasons for the secretion of sweat from the body. Some of the factors are related to the thermoregulatory system, but many others are preconditioned by life style, such as the following:

- Sweating as a part of thermoregulation due to
  - Stimulation of the skin temperature sensor.
  - Raising the temperature of the blood stream.
- Physically preconditioned sweating because of
  - Psychological stress.
  - Emotional stress.

- Pharmacologically determined sweating due to
  - Presence of nicotine in the blood.
  - Adrenaline flux.
- Specific reflex sweating due to
  - Local pain.
  - Local blood pressure rise.
  - Gustatory stimulation.

It is known that the high relative humidity causes a negative feeling in people and they assess the thermal comfort as unsatisfactory. Therefore, sweat appearance and increased humidity in the surgical room are essential for the thermophysiological comfort of the surgeons. Nocker (2011) concluded that the creation of a microenvironment with low relative humidity is even more important than the temperature decrease. One way of doing this is the use of clothing with good ventilation abilities, fast removal of water vapor, and avoidance of clothing that lead to condensation of water vapor and liquid appearance on the skin.

In this sense, surgical clothes must be treated as a system of textiles, among which air layers that need to be effectively transported to the environment are kept.

If the macrostructure, used for the manufacture of surgical clothing, has pore sizes smaller than the size of microbes, the microbes cannot penetrate into the body of a healthy person. But it is not always possible to design a textile with a desired pore size that prevents this penetration, particularly when referring to viruses of a micro-size. The application of a non-permeable coating, however, deteriorates the thermophysiological comfort of the person, as the removal of the latent heat flow from the body becomes impossible (see Chapter 2).

The absence of thermophysiological comfort cannot be a serious problem for short operations, but it is essential for complex and lengthy surgical interventions. Therefore, intensive research is performed for development of such impermeable, but *breathable* coatings that impede fluids such as blood and serum, but allow the air flow through the surgical clothing. This requirement is valid even to the *classical* woven macrostructures for surgical clothes with very high densities: their air permeability should be guaranteed.

The requirements to the bottom of the sleeves and the front of surgical gowns are particularly high, because they are the nearest to the patient's wound area and are at greater risk of penetration of infection: from and to the patient. This feature is taken into account in the production of *clean air suits*: a special kind of medical clothing, which reduces the particles emitted by the human body by additional filtering effect. To reduce the cost of production, *clean* parts like collars, sleeves, and others may be included in the protective clothing of the surgeons instead of a complete suit. Clean air suits are used by medical staff instead of normal clothing or under the surgical gown.

8.2.2 Anesthesiologists and Nurses

The rest of the medical staff in the operating room, that is, the anesthesiologist, anesthesia and surgical nurses, are away from the *operational zone*, and unlike the surgeon and the patient, they do more active movements. They are closest to the state of thermal neutrality, if the parameters of the indoor environment (temperature, humidity, etc.) are in accordance with the requirements of the thermal comfort standard.

The ergonomic requirements for the clothing of nurses are much higher than the requirements for providing thermophysiological comfort: they must be non-transparent, soft in touch, in bright colors, and nice design so as to create the required aesthetic comfort of patients and visitors in the health-care institutions (Walz 2011). At the same time, clothing of nurses has to provide freedom of movement during manipulations and especially during the transportation of patients, which is combined with physical efforts and can result in thermal discomfort.

8.2.3 Patient

The patient in the operating room is in a state of immobility, combined with anesthesia and (often) lack of awareness. The body is naked under the drape and, although also laying in the *hot* area under the lights (zone 1), is threatened by hypothermia due to intervention, blood loss, low blood pressure, stress, stiffness, and inability of the thermoregulatory system to respond to the decreasing temperature of the core body. Therefore, the patient is at risk of hypothermia in the same environment, in which surgeons are in thermal discomfort due to the increased air temperature.

During a surgery, the thermoregulatory system of the body is insufficient to maintain the temperature of the internal organs in the range of 35.5°C–37°C due to the extreme conditions. Children and the elderly people are especially at a risk. Children's thermoregulatory system cannot use muscles contraction to generate heat. At the same time, the surface of the skin through which the body exchanges heat with the environment is relatively large compared to body weight. The problems of elderly people arise from diminished metabolic reactions, and age-related diseases further enhance the risk of hypothermia.

Different means, active and passive, are used for warming of patients during surgery and in the intensive care unit. Active devices include mainly blankets with hot air or electric blankets. Their disadvantage is that they are not flexible enough, are not air permeable, and often hinder the work of surgeons and nurses.

Woven drapes and covers, including laminated, three-dimensional woven fabrics and heat insulating nonwoven textiles, are the passive devices with greatest application against hypothermia. Their main task is to provide high thermal insulation with maximum breathability. Therefore, the air permeability of these products is of particular importance. Each of these structures has its great advantage over the other: woven textiles are reusable and subject to disinfection, as well as they are relatively light at high density (warp and weft densities). Knitted fabrics have the highest potential for thermal insulation, while the nonwoven products assure the highest protection against infections. The disadvantage of the woven textiles, related with the insufficient thermal insulation for the patient in the operating room, can be overcome by using microfibers or several layers from one and the same drape (Möhring et al. 2011).

8.3 Properties of Single Woven Structures for Surgical Clothes and Drapes

The main task in this chapter was to examine systematically the performance properties of woven macrostructures used for surgical clothes and medical drapes, which properties are associated with thermophysiological comfort. Structural, geometric, and mass characteristics were determined together with moisture transfer and air permeability of single layer fabric.

Special attention was paid to the assessment of the transfer of air through systems of layers: two identical layers, three identical layers, and two different layers. The results were analyzed and compared with theoretical results from prediction of air permeability through two or more layers according to the model of Clayton (1935).

New theoretical models for systems of layers from woven textiles were derived based on experimental data for the air permeability of a single layer and systems of layers. The model allows to predict the air permeability coefficient of a system of layers when the air permeability coefficient of the single layer (single layers) is known. Verification of the model was done in Chapter 10 with a new set of woven macrostructures.

Nine woven macrostructures used for the production of surgical clothes and medical drapes were tested for determination of their structural, geometric, and mass parameters. The fabric weight, warp and weft densities, and thickness were determined on the basis of 10 single measurements. The linear density of warp and weft threads and their twist were determined on the basis of 20 single measurements. The methods described in Chapter 5 were used for the measurements. The average results are summarized in Table 8.1. (The samples were arranged by increasing of the fabric weight.)

TABLE 8.1

Data for the Woven Macrostructures for Surgical Clothes and Drapes

Sample	Composition	Fabric Weight, g/m²	Warp Density, threads/dm	Weft Density, threads/dm	Linear Density, tex		Twist, m⁻¹		Weave	Thickness, mm
					Warp	Weft	Warp	Weft		
1	CO/PES 33/67%	113	270	410	15	12	1000	999.4	Plain	0.237
2	CO/PES 33/67%	116	260	430	14	16	805.5	908.8	Twill 1/2Z	0.251
3	CO/PES 33/67%	119	200	280	16	15	888.3	837.7	Twill 1/2Z	0.267
4	PES 100%	122	390	400	16	15	896.1	863.8	Plain	0.25
5	PES 100%	126	290	340	16	14	938.8	871.1	Plain	0.26
6	CO 100%	160	260	300	30	28	795.5	833.8	Plain	0.43
7	CO 100%	165	260	310	26	20	845	349.4	Twill 2/2Z	0.355
8	CO/PES 33/67%	175	240	350	27	28	363.3	412.7	Twill 2/2Z	0.42
9	CO/PES 33/67%	185	250	270	34	37	821.6	668.3	Plain	0.421

8.4 Experimental Assessment of the Air Permeability of a Single Layer

The air permeability of the tested samples was determined as described in Chapter 5. Twenty measurements were made for each sample, and Table 8.2 shows the results for the air permeability coefficient B_p. The table also includes results from measurements of warp and weft cover factor, fabric cover factor, and areal porosity.

Figures 8.1 through 8.5 illustrate the dependence of the air permeability of a single layer on the fabric weight, thickness, warp linear filling, weft linear filling, and areal porosity. The study and related analysis are similar to the study on the woven macrostructures for clothing and bedding presented in Chapter 7.

TABLE 8.2

Other Structural Parameters and Air Permeability Coefficient

Sample	Linear Cover Factor E, %		Fabric Cover Factor E_s, %	Fabric Areal Porosity V_s, %	Air Permeability Coefficient B_p, m/s
	Warp	Weft			
1	42.36	57.53	75.52	24.48	0.158
2	38.7	70.54	81.94	18.06	0.180
3	32.41	43.93	62.10	37.90	0.147
4	63.19	62.75	86.29	13.71	0.099
5	46.25	52.08	74.24	25.76	0.104
6	57.2	64.99	85.01	14.99	0.143
7	54.01	56.16	79.84	20.16	0.119
8	50.33	75.02	87.59	12.41	0.175
9	59.05	66.79	81.74	18.26	0.098

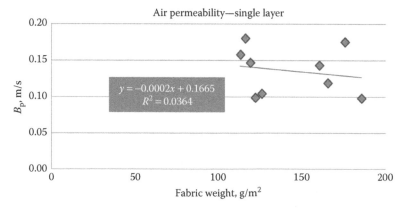

FIGURE 8.1

Air permeability as a function of the fabric weight—single layer.

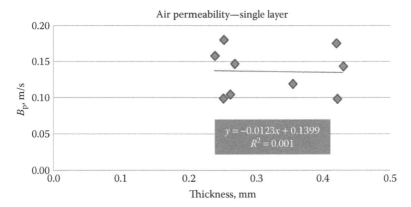

FIGURE 8.2
Air permeability as a function of the thickness—single layer.

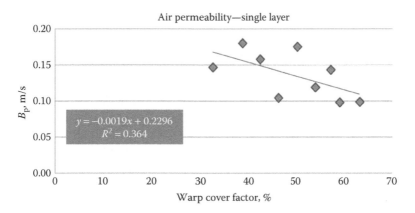

FIGURE 8.3
Air permeability as a function of the warp cover factor—single layer.

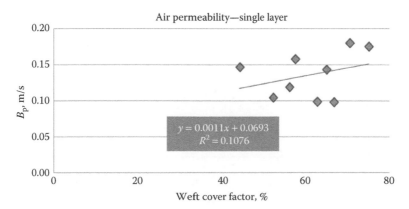

FIGURE 8.4
Air permeability as a function of the weft cover factor—single layer.

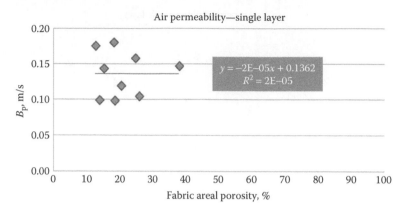

FIGURE 8.5
Air permeability as a function of the fabric areal porosity—single layer.

The results showed, however, that there were no regression equations between the air permeability coefficient and tested characteristics of the macrostructure with a correlation coefficient $R^2 > 0.5$. The dispersion of the results was apparent. Provided that the selected woven samples were united only by their *general application*, the question arose whether the choice of other samples or eliminating part of the already tested samples would change the picture of the dependences.

Based on the results in Figure 8.1, three of the nine macrostructures, where the deviations were among the largest, namely, samples 4, 5, and 8, were eliminated. The result of the relationship between the fabric weight and the air permeability coefficient B_p is shown in Figure 8.6. The much better grouping of the values was not only obvious, but it was reflected by the resulting statistical correlation with higher correlation coefficient ($R^2 = 0.79$):

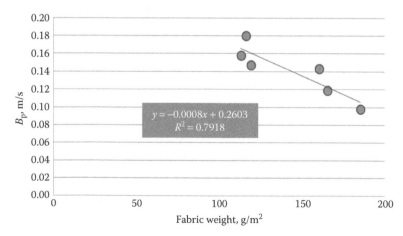

FIGURE 8.6
Air permeability as a function of the fabric weight—six samples.

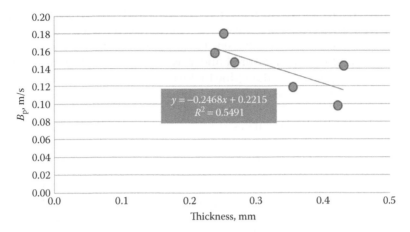

FIGURE 8.7
Air permeability as a function of the thickness—six samples.

$$B_p = -0.0008 m_s + 0.2603 \tag{8.1}$$

The derived regression equations between the same quantities in Chapter 7 had a correlation coefficient $R^2 = 0.89$.

Figure 8.7 presents the results of the relationship between B_p and the thickness for the six samples. The correlation coefficient of the derived linear statistical dependence increased again and reached $R^2 = 0.55$. The correlation coefficient of the similar regression equation in Chapter 7 was $R^2 = 0.5$.

8.5 Experimental Assessment of the Air Permeability of Multilayer Systems

The number of textile layers in the clothing is essential, because a critical aspect for the insulation abilities of each clothing ensemble is the layer/ layers of air, retained between the system of textile layers. Since air transfers heat more slowly than the textile fibers, two or more layers of fabric for surgical clothes and drapes can provide better comfort for the patient or impair the comfort of the surgeon (Fourt and Hollies 1970). In all cases, the macrostructures that have lower air permeability will ventilate the body less.

In this part of the study, the air permeability of systems of two and three woven macrostructures was experimentally determined as follows:

- System of two identical layers
- System of three identical layers
- System of two different layers

Warp and weft threads of the layers coincided and a polypropylene mesh was placed between any two layers. The mesh had a thickness of 0.13 mm and eliminated the contact between the samples during the measurement. Its air permeability was measured and it was found that the mesh had no resistance to the passing airflow (due to the large pores between the threads).

The measuring area was 10 cm², and the pressure difference set between the two sides of the textile systems was 100 Pa. Twenty single measurements were done for each individual system.

8.5.1 Theoretical Model of Clayton

Each textile layer, participating in the system, has its own air permeability, which can be measured. The combination of two or more layers, even of the same macrostructure, changes the flow rate of the passing fluid. The main reason for this is the reduced size of the spaces between the threads as a result of the coincidence of the threads of one layer with the pores of the other.

The main aim of a theoretical model for determination of the air permeability of a system of two or more layers is to predict the air permeability of the system based on the knowledge of the air permeability of the single layer. Clayton (1935) proposed such a formula for calculation of the air permeability AP_N of a system of N layers:

$$AP_N = \left(\frac{1}{AP_{11}} + \frac{1}{AP_{12}} + \ldots \frac{1}{AP_{1N}} \right)^{-1} \tag{8.2}$$

where:
AP_{11} to AP_{1N} is the air permeability coefficient of the single layers, m/s

Clayton's model did not make a difference between the types of the layers and their number. It was applied for theoretical evaluation of the air permeability of the studied systems of woven textiles and the results of the predictions were compared with the experimental results obtained.

8.5.2 Air Permeability of a System of Two Identical Layers

Five of the samples with light to medium fabric weight were tested: samples 1 and 2 (light) and samples 6, 7, and 8 (medium). The tested system included two layers of each sample.

Figure 8.8 illustrates the results for the air permeability coefficient B_p of a single layer of the examined woven macrostructures (see Table 8.2). Table 8.3 and Figure 8.9 show the results from the measurement of the air permeability of a system of two identical layers. The theoretical values of the air permeability coefficient of a system of two identical layers, based on Clayton's method, were also calculated.

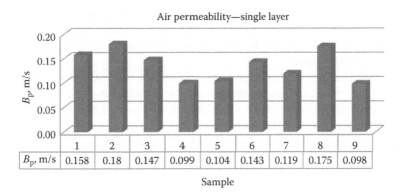

FIGURE 8.8
Air permeability of a single layer.

TABLE 8.3

Air Permeability of a System of Two Identical Layers

| Sample | Air Permeability Coefficient B_p, m/s | | |
	Model of Clayton	Experimental Values	Relative Error, %
1	0.097	0.091	−15.32
2	0.090	0.099	−10.04
6	0.071	0.077	−7.71
7	0.060	0.064	−7.51
8	0.088	0.100	−14.00

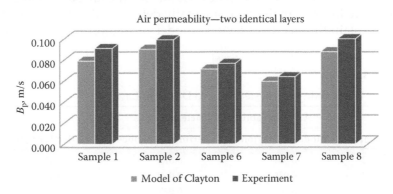

FIGURE 8.9
Air permeability of a system of two identical layers.

The reduction in the air permeability of the system, in comparison with the transport of air through a single layer, was obvious. The decrease of B_p was from 42.4% (for sample 1) to 46.2% (for sample 7), which determined a relatively narrow range.

The comparison between experimental and theoretical values of B_p of the system of two identical layers confirmed the conclusions of Lord (1959) that for textiles with high densities, Clayton's model gave lower theoretical values than the experimentally measured. The relative error between theoretical and experimental results (calculated as mentioned in Chapter 5) is also listed in Table 8.3.

8.5.3 Air Permeability of a System of Two Different Layers

Combinations between sample 1 and samples 2, 3, 5, 6, and 7 were studied. Sample 1 was always a top layer so as to eliminate the expected effect of the order of layering of the samples in the system. Clayton's model was used again for theoretical calculation of the air permeability coefficient of the system.

The theoretical and experimental results for B_p are summarized in Table 8.4 and compared in Figure 8.10. In this case, the model of Clayton gave higher values than the experiment, which is evident from the calculation of the relative error in Table 8.4.

8.5.4 Air Permeability of a System of Three Identical Layers

Samples 1, 2, 6, 7, and 8 were again subject of the study, but the systems included three layers of each sample. The results of the comparison between the theoretical and experimental results are listed in Table 8.5 and visualized in Figure 8.11.

Similar to Figure 8.9, the theoretical results for the air permeability coefficient B_p are lower than the experimentally determined. It is difficult, however, to explain the fact that Clayton's model gave lower predicted value of

TABLE 8.4

Air Permeability of a System of Two Different Layers

| Sample | Air Permeability Coefficient B_p, m/s | | Relative Error, % |
	Model of Clayton	Experimental Values	
1 + 2	0.084	0.075	10.87
1 + 3	0.076	0.068	10.80
1 + 5	0.063	0.049	21.16
1 + 6	0.075	0.065	12.97
1 + 7	0.068	0.061	10.25

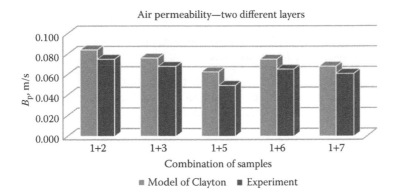

FIGURE 8.10
Air permeability of a system of two different layers.

TABLE 8.5

Air Permeability of a System of Three Identical Layers

Sample	Air Permeability Coefficient B_p, m/s		
	Model of Clayton	Experimental Values	Relative Error, %
1	0.053	0.061	−16.41
2	0.060	0.064	−6.56
6	0.048	0.057	−19.35
7	0.040	0.047	−19.47
8	0.058	0.063	−7.73

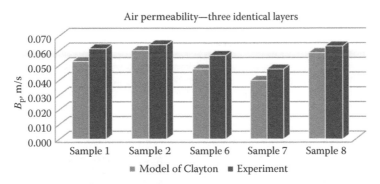

FIGURE 8.11
Air permeability of a system of three identical layers.

the coefficient of air permeability for a system of two or three identical layers, but for a system of two different layers (Figure 8.10) the theoretical results were higher for all measured systems. For a more accurate assessment of the results, the relative error between the experimental and the theoretical values of B_p was calculated and listed in Table 8.5.

8.5.5 Regression Models for the Air Permeability of Systems of Layers

8.5.5.1 Regression Models for the Air Permeability of a System of Identical Layers

The experimental results of measuring the air permeability of a system of two or three layers of one and the same woven macrostructure were used. The change of the air permeability coefficient B_p of the systems of layers was built as a function of the air permeability of the single layer B_{p1} and presented in Figure 8.12. The regression equations for the systems of two and three layers were of a linear type, with a high correlation coefficient, as follows:

- For two identical layers ($R^2 = 0.98$):

$$B_{p2L} = 0.62B_{p1} - 0.0092 \qquad (8.3)$$

where:

   B_{p2L} is the air permeability coefficient of a system of two identical layers

- For three identical layers ($R^2 = 0.95$):

$$B_{p3L} = 0.27B_{p1} + 0.0172 \qquad (8.4)$$

where:

   B_{p3L} is the air permeability coefficient of a system of three identical layers

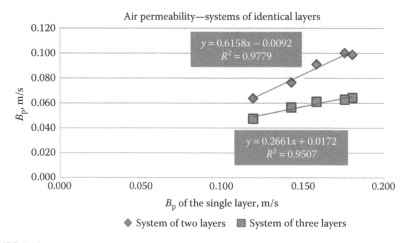

FIGURE 8.12
Air permeability of systems of identical layers as a function of the air permeability of the single layer.

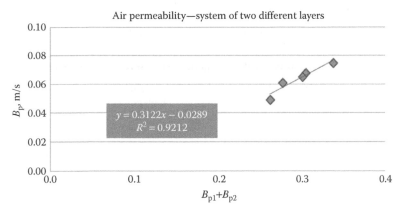

FIGURE 8.13
Air permeability of a system of two different layers as a function of the air permeability of the single layers.

8.5.5.2 Regression Models for the Air Permeability of a System of Different Layers

Due to the impossibility to apply the same approach for the system of different layers (no specific single layer), a relationship between the air permeability of the system and the sum of the air permeability of the participating single layers ($B_{p1} + B_{p2}$) was searched. The results are presented in Figure 8.13. A linear regression equation was derived with a correlation coefficient $R^2 = 0.92$:

$$B_{p2 \neq L} = 0.3122\left(B_{p1} + B_{p2}\right) - 0.0289 \tag{8.5}$$

where:

$B_{p2 \neq L}$ is the air permeability coefficient of the system of two different layers

8.6 Theoretical Models for the Air Permeability of Systems of Layers

8.6.1 Theoretical Model for the Air Permeability of a System of Identical Layers

Table 8.6 summarizes the results for the air permeability of the single layers and for the systems of two and three identical layers.

The dependence of the air permeability of the systems of two and three layers from the air permeability of the single layer can be generally presented as follows:

TABLE 8.6

Air Permeability of Systems of Identical Layers—Experimental
Results and Coefficients of Proportionality

	Air Permeability Coefficient B_p, m/s			Coefficients	
Sample	One Layer	Two Identical Layers	Three Identical Layers	k_1	k_2
1	0.158	0.091	0.061	0.58	0.39
2	0.180	0.099	0.064	0.55	0.36
6	0.143	0.077	0.057	0.54	0.40
7	0.119	0.064	0.047	0.54	0.40
8	0.175	0.100	0.063	0.57	0.36

$$B_{p2A} = k_1 B_{p1}$$
$$B_{p3A} = k_2 B_{p1}$$
(8.6)

where:

k_1 and k_2 are coefficients of proportionality

The calculated coefficients k_1 and k_2 for the investigated samples are also presented in Table 8.6. The establishment of a regression model between the number of layers in the system N and the average of the coefficients of proportionality k_1 and k_2 is presented in Figure 8.14.

As a result, a new model, for calculation of the coefficient of air permeability of the system of identical layers as a function of the coefficient of air permeability of the single layer B_{p1}, was derived (Equation 8.7):

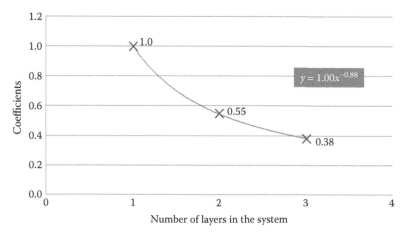

FIGURE 8.14
Determination of the coefficients k_1 and k_2.

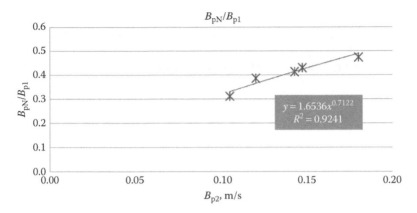

FIGURE 8.15
Determination of the coefficient k.

$$B_{pN} = B_{p1}N^{-0.88} \tag{8.7}$$

8.6.2 Theoretical Model for the Air Permeability of a System of Different Layers

By analogy with Equation 8.6, the air permeability of the system of two different layers can be presented as follows:

$$B_{pN} = kB_{p1}B_{p2} \tag{8.8}$$

Regression analysis was again used for determination of the coefficient k, and the result of the obtained regression equation is shown in Figure 8.15.

As a result, the final equation, which is a model for prediction of the air permeability of a system of two different layers, was derived (Equation 8.9):

$$B_{pN} = 1.65B_{p1}B_{p2}^{0.71} \tag{8.9}$$

8.7 Experimental Assessment of the Moisture Transport

Six of the selected woven macrostructures were tested, namely, samples 2, 3, 6, 7, 8, and 9. The method described in Chapter 5 was used; five samples were tested from each macrostructure. Table 8.7 presents the results for the evaluated characteristics of the moisture transport through the samples (SDL Atlas 2010).

In accordance with the wetting time, the samples wet fast or very fast, and only samples 6, 8, and 9 were characterized as slowly wetting, taking into account the results for the wetting time of the top surface (WTT).

TABLE 8.7

Moisture Transport through the Macrostructures for Surgical Clothes and Medical Drapes

Sample	WTT, s	WTB, s	TAR, %/s	BAR, %/s	TSS, mm/s	BSS, mm/s	R_{index} mm²/s	OMMC
2	3.09	2.44	4.79	75.57	1.21	2.51	984.60	0.807
3	6.09	2.67	10.49	59.60	1.74	4.07	803.47	0.872
6	45.98	3.37	8.79	53.01	0.78	3.91	994.67	0.847
7	12.56	2.76	25.38	53.52	0.85	3.73	995.10	0.849
8	51.73	3.47	14.37	22.06	1.63	1.64	387.78	0.607
9	31.17	4.31	28.76	25.47	0.29	2.18	930.94	0.641

The top absorption ratio (TAR) determined the samples as slowly or very slowly wetting (TAR < 30), while the bottom absorption ratio (BAR) assessed them as fast to moderately fast wetting. The TSS characteristic also assessed the samples as very slowly to slowly wetting (TSS < 2), but the similar characteristic for the bottom side (BSS) determined that the samples are fast wetting. The results for the accumulative one-way transport index characterized the macrostructures as perfectly providing moisture from top to bottom surface. Sample 8 is the only exception, as its transport abilities are *very good* (R_{index} < 400).

The overall moisture management capacity (OMMC), which reflects the ability of the samples to transfer liquids in both vertical and horizontal plane, showed that the transport abilities of samples 8 and 9 are *very good*, and all other samples transport moisture perfectly.

Similar to the analysis performed in Chapter 7, here the effect of fabric weight, thickness, and areal porosity on moisture transfer was assessed. For this purpose, graphical dependencies between the three characteristics of the woven macrostructures and the eight characteristics of the moisture transfer were built. Regression equations of linear type were derived and their correlation coefficients were summarized in Table 8.8.

TABLE 8.8

Summary of the Correlation Coefficients for the Linear Relationship between the Characteristics of the Macrostructure and the Characteristics for Moisture Transfer

	Correlation Coefficient R^2							
	WTT, s	WTB, s	TSS, mm/s	BSS, mm/s	TAR, %/s	BAR, %/s	R_{index} mm²/s	OMMC
Fabric weight	0.54	0.69	0.32	0.16	0.55	0.75	0.06	0.43
Thickness	0.39	0.74	0.22	0.23	0.68	0.88	0.12	0.56
Areal porosity	0.43	0.17	0.17	0.35	0.01	0.15	0.02	0.29

8.7.1 Effect of Fabric Weight

The results showed that regression equations for the effect of the fabric weight on the moisture transfer through the macrostructures with higher correlation coefficients are obtained, compared with the models in Chapter 7. Four of the eight equations have a correlation coefficient $R^2 > 0.5$ as follows:

- *WTT*: Figure 8.16, with $R^2 = 0.54$

$$\text{WTT} = 0.5293m_s - 56.057 \tag{8.10}$$

- *WTB (wetting time bottom)*: Figure 8.17, with $R^2 = 0.69$

$$\text{WTB} = 0.0198m_s + 0.1373 \tag{8.11}$$

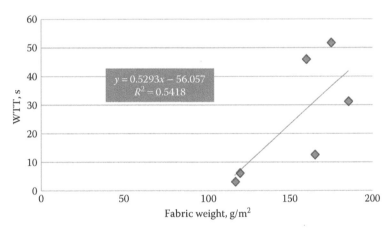

FIGURE 8.16
Influence of the fabric weight on the wetting time top.

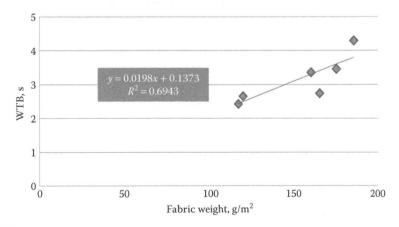

FIGURE 8.17
Influence of the fabric weight on the wetting time bottom.

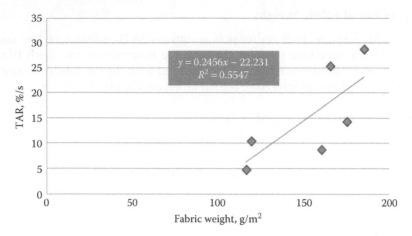

FIGURE 8.18
Influence of the fabric weight on the top absorption ratio.

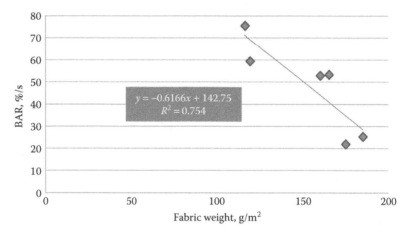

FIGURE 8.19
Influence of the fabric weight on the bottom absorption ratio.

- *TAR*: Figure 8.18 with $R^2 = 0.55$

$$\text{TAR} = 0.2456m_s - 22.231 \tag{8.12}$$

- *BAR*: Figure 8.19 with $R^2 = 0.75$

$$\text{BAR} = -0.6166m_s + 142.75 \tag{8.13}$$

8.7.2 Effect of Thickness

The regression equations for the dependence of three of the studied characteristics for moisture transfer on the thickness of the macrostructures had a correlation coefficient $R^2 > 0.5$, similar to the results in Chapter 7, where only

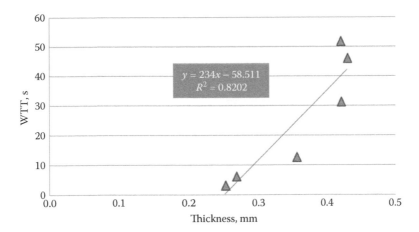

FIGURE 8.20
Influence of the thickness on the wetting time top.

two of the eight studied characteristics had a statistically proven dependence on the thickness (with a correlation coefficient $R^2 > 0.5$).

The highest correlation coefficient was that of the dependence between the thickness and the WTT surface of the samples ($R^2 = 0.82$). The derived regression equation (Figure 8.20) was

$$WTT = 234\delta - 58.511 \tag{8.14}$$

The regression equation between the thickness and the time for wetting of the bottom surface (WTB) had a correlation coefficient $R^2 = 0.68$ (Figure 8.21):

$$WTB = 7.0161\delta + 0.6629 \tag{8.15}$$

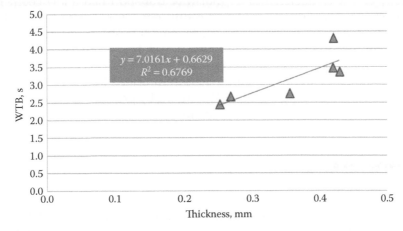

FIGURE 8.21
Influence of the thickness on the wetting time bottom.

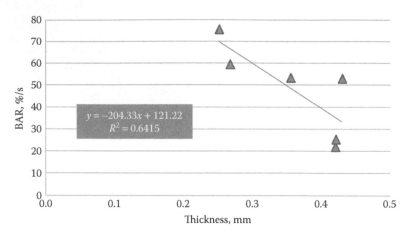

FIGURE 8.22
Influence of the thickness on the bottom absorption ratio.

Similar was the regression coefficient of the dependence between the thickness and the BAR ($R^2 = 0.64$) (Figure 8.22):

$$BAR = -204.33\delta + 121.22 \tag{8.16}$$

The highest correlation coefficients were obtained for the regression equations between the thickness and both the bottom spreading speed (BSS) ($R^2 = 0.64$) and the OMMC ($R^2 = 0.56$).

8.7.3 Effect of Areal Porosity

The analysis of the results in Table 8.8 showed that, similar to the results in Chapter 7, the derived regression equations for the effect of the fabric areal porosity on the characteristics for moisture transport had a low coefficient of correlation. The highest correlation coefficient was obtained for the effect of the areal porosity on the WTT surface of the woven macrostructures ($R^2 = 0.43$). The calculation of the linear correlation coefficient of all other tested characteristics of moisture transfer through the macrostructures and the examination with the student's criterion indicated that only two characteristics, BSS and the OMMC, showed statistically proven dependence on the areal porosity of the woven macrostructure.

8.8 Summary

An analysis and evaluation of the thermophysiological comfort of the occupants of a surgical room in terms of their roles and clothing (textiles), which are in contact with the human body, were presented.

A systematic experimental study on the effect of the parameters of woven macrostructures for surgical clothing and medical drapes on the transfer of air and moisture was performed. The obtained results supplemented the results of Chapter 7 with a range of woven structures with different applications.

The new facts, established in Chapter 7, on the presence of a statistical relationship between characteristics of the woven macrostructures—fabric weight and thickness—and the transfer of air and moisture were confirmed once again.

A systematic experimental study on the air permeability of three types of systems of layers of woven macrostructures was performed, including systems of two and three identical macrostructures and two different macrostructures. Regression analysis was used to derive dependences on the effect of the number and type of the textile layers on the air transfer in through-thickness direction of the system.

It was found that the air permeability coefficient of the system of layers decreased with the increase in the number of layers. The decrease was as an average 44% for a system with two identical layers and 63% for a system with three identical layers, compared with the air permeability of the single layer. It was found that the decrease in the air permeability coefficient was higher for a system of two different layers in comparison with a system of two identical layers of any of the macrostructures.

A new model for calculating the air permeability of systems of two and three identical layers of woven macrostructures was derived. A new model for prediction of the air permeability of a system of two different layers of woven macrostructures was also derived. The two models require knowledge on the air permeability of the single layers, included in the systems.

9

Experimental Investigation of the Macrostructure of Upholstery Textiles

9.1 State of the Art of the Problem

In its annual document for standards and guidance to producers and consumers of furniture fabrics (woven and knitted), Joint Industry Fabric Standards Committee (1994) listed a series of standards and requirements, but none of them concerned thermophysiological comfort provided by this type of textiles. Similar document, issued in 2009, again contained no such requirement (Anon 1994).

Upholstery textiles are the most outer layer of simpler or more complex furniture systems, which may include hidden textile layers as well, usually nonwoven macrostructures. The classical design of furniture (sofas, mattresses) comprises three layers (Guillaume et al. 2012):

- Upholstery textile is composed of natural fibers and materials (cotton, linen, wool, leather) and man-made fibers (polyester, polyacrylonitrile [PAN], polyvinyl chloride [PVC]), as well as mixtures of natural and chemical fibers.
- Intermediate layer is made of latex, polyester, or polyurethane foam.
- Internal layer is made of aramid or glass fibers, which should play the role of a flame retardant.

Though being the outermost layer of many furniture systems, the upholstery textiles are very rarely a subject of comments and even less of a research that address the topic of the thermophysiological comfort of the occupants in the indoor environment. In most cases, other properties of upholstery textiles, not less important, are studied, being associated with the particular operation of the indoor environment and/or contact with the human body.

The analysis of the literature showed that a series of important performance properties of upholstery textiles were subject of research. The topics of the studies, though their review does not claim to be exhaustive, can be summarized as follows:

- *Fire resistance*: Donaldson et al. (1979, 1981), Travers and Olsen (1982), Hettich (1984), Gandhi and Spivak (1984), Harper et al. (1986), Ihrig et al. (1986), Kotresh (1996), Takigami et al. (2009)
- *Color fastness*: Warfield (1987), Alpay et al. (2005), Akgun et al. (2010)
- *Resistance to UV rays*: Wagner et al. (1985), Kajiwara et al. (2013)
- *Physical and mechanical properties (pilling, strength and elongation, seam strength, touch, etc.)*: Wilson and Laing (1995), Baltakytė and Petrulytė (2008), Pamuk and Çeken (2009), Djonov and Van Leeuwen (2011), Bilisik et al. (2011)
- *Abrasion resistance*: Bilisik and Yolacan (2009), Jerkovic et al. (2010), Akgun et al. (2010)

The need to assess the behavior of the upholstery textiles in the thermophysiological aspect of the comfort was discussed by Habboub (2003), mentioning the particular case of the use of office chairs. In fact, the author considered the thermal comfort of the indoor environment, but not thermophysiological comfort. He demonstrated the importance of the structure of the outer layer of the chairs for the heat transfer, measured by thermal imaging camera. The author noted the insufficient knowledge on upholstery textiles at the expense of other textiles, that is, for clothes, in terms of their impact on the comfort of the individual in the indoor environment.

The importance of thermophysiological comfort provided by the upholstery textiles is particularly relevant for specific applications: in wheel chairs for disabled, mattresses for infants or bedridden, that is, when a prolonged contact with the human body is foreseen. The same is true for situations that go beyond the case of the built environment, but are again associated with specific indoor environment: in vehicles, especially when it comes to a long journey.

The study of Ilce and Cayir (2010), dedicated to consumer preferences in the design of the seats in the car, found that a total of 66.8% of respondents preferred textiles on seats (31.4%) or a combination of textiles and leather (33.4%). The main reason for that was the better ability of the textiles to transfer the heat from the body without causing sweating in the contact areas with the seat. The same conclusion was drawn in the work of Bartels (2003), where covers for seats in an airplane were investigated: textiles provided better conditions in terms of the local increase in temperature of the body and the occurrence of thermal discomfort compared to leather seats.

In general, the research on upholstery textiles mostly concerns the textile layer or the seat systems used in vehicles, as well as the thermal comfort in cars and airplanes (more rarely in buses and trains): Faris (1995), Schacher and Adolphe (1997), Musat and Helerea (2009), Jerkovic et al. (2010). One reason for that is economical—the companies dealing with the development and production of cars and airplanes devote significant resources to ensure the comfort of their clients and fund research in this area. On the other hand, the immobility of passengers for hours in planes or cars preconditions such studies in a much greater degree than investigations on upholstery textiles and furniture systems for offices or dwellings (the furniture for disabled people, bedridden, or babies are again an exception).

The research works of Warfield (1987), Gandhi and Spivak (1994), Snycerski and Frontczak-Wasiak (2002), Vlaović and Župčić (2012), and Licina et al. (2014) are representative of a number of studies on upholstery textiles and prove the already drawn conclusion: all they have been devoted to the assessment of the furniture system as a whole, not just the outermost textile layer.

Snycerski and Frontczak-Wasiak (2002) studied the transport of moisture through three types of furniture systems, in which the upholstery textile (knitted, with a pile) was only the upper part of the system. A positive feature of the study was that the experimental setup, designed specifically for the assessment of moisture transfer through the system, was described in detail. However, the used in the furniture system layers (including the textiles) were described only as material and thickness, which did not allow thorough analysis of the influence of the textile layer on the whole system to be made.

An important conclusion by Snycerski and Frontczak-Wasiak (2002) was that the presence of air both inside the pores of the fabric and between the layers of the system affected the transfer of moisture in the tested furniture systems. They also commented the need of the textile layer, which is in contact with the human body, to be hydrophobic, in order to provide a feeling of dryness. Thus, a better comfort could be assured at the expense of the intermediate layer that must be hydrophilic and absorb the unwanted moisture away from the body. Such a conclusion was made later by Mukhopadhyay and Midha (2008).

In fact, the conclusion of Snycerski and Frontczak-Wasiak (2002) on the impact of air is not anything, but a replica of the conclusions of the introductory presentation of Schiefer (1943) on the annual meeting of the Textile Research Institute. Even then, Schiefer (1943) noted the impact of air on the water vapor transport. Though the author didn't define and analyze the problem, it concerned the boundary layer, which lies on the border *solid body–surrounding air* in any surface. Schiefer (1943) noted that the human body constantly released *invisible sweat*, which varied from 20 g/m²h water to 1000 g/m²h and even more, depending on the conditions. The amount, however, which could be evaporated and therefore effectively be involved in the cooling of the body, was significantly less. From the point of view of current knowledge in the field, it is clear that the air in the textile layer (textile macrostructure) and between the textile layers in the system, as well as the boundary layer itself, is among the reasons for insufficient transfer of water vapor.

Laconic in content, the report of Schiefer (1943) contained a particularly important conclusion: the vapor permeability of textiles did not depend on their porosity, but on the detention or transfer of air within and between the textile layers. Therefore, the author recommended to evaluate not the vapour permeability, but the air permeability of textiles (in that particular case—military uniforms for use in the jungle).

In this sense, it can be concluded that the measurement of the vapor permeability of a single textile layer (macrostructure) is not objective and sufficient enough to assess the effect of the water vapor transfer on the thermophysiological comfort of the individual, when a layer of air exists between the human body and the textile layer. The reason is that the vapor transfer through the textile layer is measured with a certain pressure gradient. These experimental conditions, however, cannot be observed in the indoor or outdoor environment. The essential factor in this case is the existence of a thin air layer, which is located between the water vapor particles on the surface of the skin and the textile layer. This means that the transfer of water vapor in the textile macrostructure is directly dependent on its air permeability.

In conclusion, the reasons for insufficient research of thermophysiological comfort provided by the upholstery textiles can be sought in two directions:

- The requirements for upholstery textiles in terms of heat and mass transfer from the human body to the environment or vice versa are not substantially different from those for textiles with other applications. Upholstery textiles must ensure that the heat generated by the body will be transmitted to the environment so that the dissipation of heat will not create discomfort for the individual. On the other hand, the upholstery textiles must ensure the transfer of heat to such a degree that it does not lead to unwanted release of sweat from the skin surface.

- Upholstery textiles are only a part of some furniture systems, which operates in a complex with the other elements of the system. It is assumed likely that the testing of the outer layer with respect to the provided thermophysiological comfort is a wasted effort if the evaluation of the system as a whole is more important.

The fact is that a separate research on transfer of air, heat, and moisture in upholstery textiles would result in one-sided and often incorrect assessment of the impact of the furniture system on the interaction of the human body with the environment. Therefore, the study in *this chapter* is limited in nature and is in favor only of the assessment of the upholstery textiles as woven macrostructures with a specific application, following the already described at the beginning of Section II tasks.

The main aim of the study on the macrostructure of upholstery textiles was to investigate the effect of the characteristics of the macrostructure on the air permeability and thermal insulation of the samples. The results obtained complemented and enriched the results from Chapters 7 and 8 with a range of woven macrostructures with different applications. Sectional air permeability was used as a characteristic of the air flow transfer in through-thickness direction together with the already applied air permeability coefficient.

9.2 Experimental Assessment of the Air Permeability

Upholstery woolen textiles with high fabric weight were selected for the purpose of the experiment in this chapter. The choice is determined by the fact that the fabrics, containing wool fibers, are at lower risks of fire. Therefore, they are preferred for upholstering furniture in the indoor environment, including vehicles (Taylor 1991; Fung 2010). Woolen textiles are determined even as fire retardant upon contact with an open flame or heat source because they have high moisture content and high flash point (570°C–600°C).

Taylor (1991), however, concluded that the mixtures of fibers or the use of different materials in the mesostructure (threads of different fibers or filaments) was more susceptible to combustion than the homogeneous materials. The same were the conclusions of Collier and Epps (1999), who found that fabrics from blends of cotton and polyester fibers burned faster than fabrics of CO 100% or polyester (PES) 100% and thoroughly analyzed the reasons for this effect.

Kotresh (1996) found in his study on 25 samples of upholstery textiles that macrostructures with greater weight burned more slowly and are more difficult to ignite compared with lighter fabrics, regardless of their composition, which was an advantage for the woolen upholstery fabrics, produced usually with high fabric weight.

The second reason for choosing woolen type of fabrics is that they absorb moisture and could even be used as compensators of the humidity in the indoor environment (IWTO 2010)—an issue that is subject of the experimental research and analysis in Chapter 11. At the same time, they ensure good thermophysiological comfort of the occupants and therefore can be used for vehicle seats, but because of the high price either mixtures of wool and man-made fibers are used or wool fabrics are applied in high-end cars (Fung and Hardcastle 2001).

The last reason that determined the choice is that wool fabrics enrich the type of the microstructures, investigated in Section II: while cotton fabrics are considered in Chapter 7 and mixtures of cotton and man-made fibers in Chapter 8, this chapter deals with woven macrostructures from mixtures of wool and man-made fibers.

Five woven macrostructures, used as upholstery textiles, were selected. The samples were analyzed for determination of their structural, geometric, and mass characteristics. All measurements were performed as described in Chapter 5. The results are summarized in Table 9.1, where the samples are arranged in accordance with their fabric weight.

Data for the yarn crimp, linear and fabric cover factor, and fabric areal porosity are presented in Table 9.2. Obviously, the studied structures were very tight, with very high fabric cover factor.

The calculation of the air permeability coefficient was done on the basis of 20 single measurements (Table 9.3). Variation coefficients and the relative standard error are also presented.

The results for the air permeability coefficient B_p of the investigated samples as a function of the fabric weight and thickness are shown in Figures 9.1 and 9.2, respectively. The derived regression equations of linear type had a correlation coefficient $R^2 > 0.5$, as follows:

- For the effect of the fabric weight (Figure 9.1) with $R^2 = 0.53$:

$$B_p = -0.0015 m_s + 0.7929 \qquad (9.1)$$

- For the effect of the thickness (Figure 9.2) with $R^2 = 0.72$:

$$B_p = -0.3071\delta + 0.5125 \qquad (9.2)$$

It was also found that due to the high values of the warp and weft densities of all samples, the values for warp and weft cover factors were distributed in a very narrow range (see Table 9.2). Therefore, the search for regression dependences between linear cover factors or fabric cover factor and the air permeability coefficient was not appropriate, though the obtained equations would have a high correlation coefficient.

Clayton (1935) introduced a new characteristic to describe the transfer of air in the transversal direction of textiles: sectional air permeability. It was defined as

$$B_s = \frac{Q}{\delta} \qquad (9.3)$$

where:
Q is the air flow rate, m³/s
δ is the thickness of the sample, m

The author claimed that the sectional air permeability B_s was a more suitable characteristic for the air flow in through-thickness direction of the textiles than the air permeability coefficient B_p. The sectional air permeability was used in later studies dedicated to the air permeability of different types of macrostructures: Kothari and Newton (1974), Natarajan (2003), and Debnath and Madhusoothanan (2011).

TABLE 9.1

Characteristics of the Upholstery Textiles

Sample	Weave	Fabric Weight, g/m²	Linear Density, tex Warp	Linear Density, tex Weft	Density, threads/dm Warp	Density, threads/dm Weft	Thickness, mm	Composition Warp	Composition Weft
1	Rib 2/2	347	160	150	150	100	0.89	WO/VI 50/50%	WO/VI 50/50%
2	Plain	362	105	125	140	140	0.7	WO/PM 50/50 WO/VI 70/30%	WO/PM 50/50% WO/VI 70/30%
3	Twill 3/2S	371	95	120	120	60	0.98	WO/VI 50/50%	WO/VI 50/50% WO/PES 45/55%
4	Twill 2/1S	410	85	147	140	140	0.84	WO/PES 45/55%	WO/PES 45/55% WO/VI 50/50%
5	Twill K3/1S	424	100	196	140	140	1.19	WO/PC 50/50	WO/VI 50/50%

TABLE 9.2

Yarn Crimp, Cover Factor, and Areal Porosity of the Samples

| Sample | Yarn Crimp, % | | Linear Cover Factor E, % | | Fabric Cover Factor E_s, % | Fabric Areal Porosity V_s, % |
	Warp	Weft	Warp E_{wa}	Weft E_{wf}		
1	4.26	12.82	99,43	50,35	99.72	0,28
2	7.36	7.23	98,85	97,16	99.97	0,03
3	5.17	8.98	84,73	36,59	90.32	9,68
4	2.00	9.42	97,16	94,57	99.85	0,15
5	2.39	8.97	99,69	92,80	99.98	0,02

TABLE 9.3

Results from the Air Permeability Measurements

Sample	Air Permeability Coefficient B_p, m/s	Variation Coefficient, %	Relative Standard Error, %
1	0.23	4.95	2.32
2	0.33	7.57	3.54
3	0.24	3.53	1.65
4	0.20	5.10	2.93
5	0.15	5.13	2.40

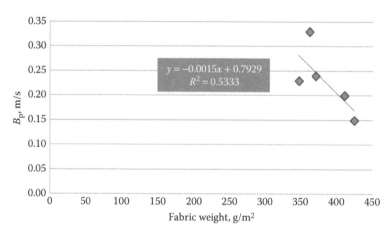

FIGURE 9.1

Air permeability as a function of the fabric weight.

The sectional air permeability B_s indicates the permeability per unit thickness of the macrostructure, which may be particularly important for textiles with a greater thickness and a high degree of linear and fabric cover factors, as it is the case of the studied upholstery textiles. Figure 9.3 presents a comparison between the coefficient of air permeability B_p, m/s, and sectional air

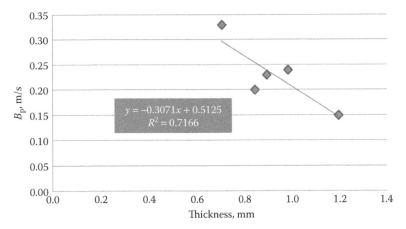

FIGURE 9.2
Air permeability as a function of the thickness.

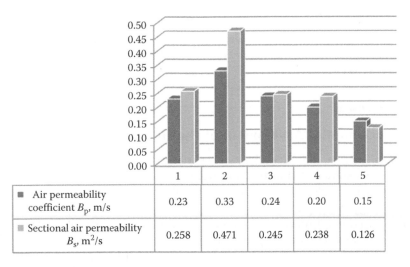

FIGURE 9.3
Comparison between the air permeability coefficient and the sectional air permeability of the macrostructure.

permeability B_s, m²/s. Obviously, a proportionality, depending on the thickness of the macrostructure, exists between the two characteristics.

A regression analysis was applied again to assess the level of the correlation between the sectional air permeability and the fabric weight. The result is shown in Figure 9.4: the correlation coefficient of the derived statistical dependence is smaller than that of Figure 9.1. The search for a relationship with the thickness is untenable.

The graph in Figure 9.4 showed that the conclusions of Clayton (1935) could not be accepted unconditionally as it was found that the sectional

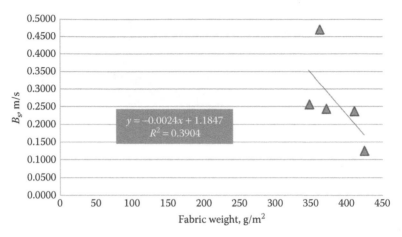

FIGURE 9.4
Sectional air permeability as a function of the fabric weight.

permeability B_s had no better statistical correlation (with higher correlation coefficient R^2) with characteristics of the woven mesostructure (fabric weight) than the air permeability coefficient B_p.

9.3 Experimental Assessment of the Heat Transfer

The results from the heat transfer through the samples of upholstery textiles—resistance to thermal conductivity R_t and thermal insulation I_{cl}— are shown in Table 9.4. The measurement was performed following the methodology described in Chapter 5.

Similar to the approach in Chapter 7 (Section II), the insulation properties of the woven samples were presented in a function of three characteristics of the macrostructure: thickness, fabric weight, and areal porosity. It was assumed that other structural indicators (i.e., warp and weft density) were implicitly reflected by the selected characteristics of the macrostructure.

TABLE 9.4

Results from the Measurement of the Thermal Insulation of the Samples

Sample	Thermal Resistance R_t, m²K/W	Thermal Insulation I_{cl}, clo
1	0.0082	0.0529
2	0.0078	0.0503
3	0.0101	0.0652
4	0.0096	0.0619
5	0.0109	0.0703

The dependence between the thermal insulation and the fabric weight of the samples is presented in Figure 9.5. The correlation coefficient of the derived linear regression equation is $R^2 = 0.63$, and the equation is

$$I_{cl} = 0.0002m_s - 0.0176 \qquad (9.4)$$

For comparison, the respective relationship in Chapter 7 had a correlation coefficient $R^2 = 0.78$. Although the value of the statistical dependence R^2 is a bit lower here, the graph of Figure 9.5 reflects the same trend—the thermal insulation increases with the increase in the fabric weight.

The change of the thermal insulation as a function of the thickness is shown in Figure 9.6. The trend expected was the thermal insulation properties of

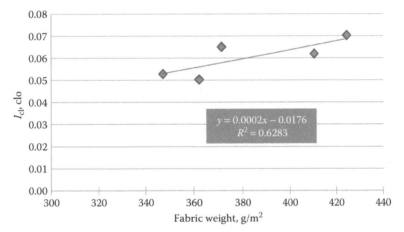

FIGURE 9.5
Thermal insulation as a function of the fabric weight of the macrostructure.

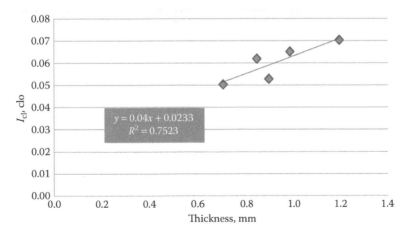

FIGURE 9.6
Thermal insulation as a function of the thickness of the macrostructure.

the samples to increase with the increment of thickness. The correlation coefficient $R^2 = 0.75$ of the derived regressions equation is very similar to the obtained regression for the macrostructures for clothing and bedding ($R^2 = 0.77$) in Chapter 7:

$$I_{cl} = 0.04\delta + 0.0233 \tag{9.5}$$

Obviously, the derived equations are valid within the scope of the tested samples.

9.4 Summary

An analysis of the role of upholstery textiles for the thermophysiological comfort of the occupants of the indoor environment was presented.

A systematic experimental study of the effect of the characteristics of the macrostructure of upholstery textiles on the transmission of air and heat was performed. The results obtained supplemented the results of Chapters 7 and 8 with a range of textiles with other application in the indoor environment.

A comparison was made between two characteristics of the air transfer in through-thickness direction: the air permeability coefficients and sectional air permeability. A positive effect of the replacement of the coefficient of air permeability in the case of woven macrostructures was not established.

It was found that for tightly woven macrostructures (with high values of warp and weft density), areal porosity/fabric cover factor area are not adequate structural characteristics to search for correlation with characteristics, describing the heat and mass transfer processes through the macrostructure.

10

Experimental Investigation of the Macrostructure of Textiles for Packing

The textiles for packaging are not a part of the indoor environment, which can be associated with the human comfort. But the polypropylene woven bags (PPWB), which are the object of the study in this chapter, are a specific product with very interesting macrostructure: these textiles are made of strips, which are not able to transfer fluids. In this context, the evaluation of the behavior of the macrostructure only binds to its porosity, without any influence of the pores in the mesostructure, which exist in the case of staple fiber yarns or polyfilaments.

Certainly, the packaging of the products is a specific item in the indoor environment: the preservation of certain goods shall be carried out in storage rooms where the temperature and humidity are controlled just as the microclimate in residential buildings. Besides, there is an intensive exchange between the room air and the stored materials. The suitability of the packaging for the storage of certain types of goods (typically food) is determined by the material from which it is made and its properties related to the transport of air, moisture, and heat. Great part of the packaging used for the transport and storage of agricultural products and food today has woven macrostructure.

This chapter presents an experimental study of the macrostructure of woven polypropylene textiles used for packaging. The reason for choosing this particular type of packaging is determined by their growing distribution and obvious advantages over woven packaging from natural materials (hemp, jute). In recent years, there is a tendency to replace paper packaging with woven polypropylene packaging. The specific air and gas permeability of the PPWB is essential for the proper storage of goods.

10.1 State of the Art of the Problem

PPWB are used for packing and storage of a large variety of products in industries such as agriculture (fertilizers, seeds, and grains), chemical, mining, building materials (sand, cement), foods (beans, rice, dried fruits, sugar, salt, coffee beans, milk powder), pet food, and coal (Angelova and Nikolova 2009).

Polypropylene items lose their advantages and deteriorate if exposed to UV, and this is one of the problems associated with their applications. Polypropylene woven products can be protected from direct sunlight both before and after weaving, but UV-resistant PPWB are more expensive. But this disadvantage is not essential if PPWB are used for storage indoors (i.e., warehouses).

The quality of the graphic is another problem with the PPWB. Paper and polyethylene are much better mediums for printing of crisp and bright graphics. However, recent achievements in the production of two- and three-layered PPWB have improved significantly the printing abilities, giving the consumers high-quality graphics at the same low production costs (Highland 1981; Kullman et al. 1981; Fleural-Lessard and Serrano 1990).

Apart from the disadvantages, which can be successfully overcome, the PPWB have several advantages over traditional paper and jute bags, which are frequently used for storage of bulky materials (Highland 1981). PPWB are lighter in weight, have higher resistance to tearing and bursting strength, and are better pest resistant. At the same time, their properties do not degrade if the bags are wet and their macrostructure is air permeable and resistant to bacteria, moisture, pollutants, oil, and mildew. Last but not least, PPWB have lower prices than paper and jute bags and can be reused.

In the recent years, a world spread tendency toward switching from paper to woven polypropylene bags for storage and transport of goods can be observed. The companies, which are using PPWB to ship their goods, report for a zero bags breaking during transportation. A year-long field study of Donahaye et al. (2007) in India showed that the breakage rate for cement PPWB was 0.04%. Laboratory tests demonstrated that the drop of paper bags from a certain height resulted in total destroying. The PPWB did not destroy from the similar height, but two or more times higher. At the same time, they usually broke along a seam, but their macrostructure remained entire (Donahaye et al. 2007).

One of the most important properties of the woven textiles for packaging is their ability to ventilate the goods and to transfer air and gases from the outside inward and from the inside outward. This particular feature of the woven macrostructures, combined with great strength, is essential for proper storage of goods. The ventilation of food, pet food, agricultural products, and others is necessary to protect them from the development of mold, bacteria growth, and moisture absorption, regardless the storage is indoors or outdoors. PPWB with sands, for example, are used as a part of construction systems for drainage; therefore, they need to be flow permeable. Because of production and storage reasons, some of the PPWB are produced with two or three layers from the same or different materials, thickness, and weave of the macrostructure (Highland 1981; Luo et al. 2001).

The storage of agricultural products in the indoor environment is a particular, but important, problem worldwide due to possible appearance of pests.

The preservation of the stored foods from pests requires fumigation (often with phosphine PH_3), which is done through the bags. Therefore, the packages must be permeable and the dense bags of polyethylene film and paper are inappropriate (Donahaye et al. 2007; Saeed et al. 2008).

The fumigation through dense packing, which does not allow the penetration of gas to all the volume of the product, can make pests resistant to this treatment, which is a concomitant defect and condition of spoilage. It should be mentioned that the woven macrostructures of natural fibers (jute, hemp) are not suitable for the storage of agricultural products due to their smaller densities and the presence of pores that are large enough to ensure the spread of harmful organisms. Therefore, polypropylene textiles are best porous textiles for manufacture of bags for storage and transportation of agricultural products, which require fumigation.

The main aim of this chapter was to study the structural and geometric characteristics and porosity of woven macrostructures, used for the production of PPWB, to complement and enrich the results from Chapters 7 through 9 with a set of woven fabrics with different applications. The objective was to determine the effect of the parameters of the woven macrostructure on the air permeability in the absence of air transfer through the mesostructure.

The air permeability of systems of two layers, positioned at an angle of 90° and 45° to each other, was also studied and used for verification of the developed new theoretical model (in Chapter 8) for air permeability of systems of identical layers.

10.2 Properties of the Macrostructures

Five woven macrostructures, used for the production of polypropylene bags, were studied. Four of them (samples 1–4) were used for direct sewing of the bags (single- or double-layered) and sample 5 was used as an internal reinforcing layer. The samples were tested to determine their structural, geometric, and mass characteristics, according to the methods described in Chapter 5.

Figure 10.1 shows the microscopic pictures (4× enlargement) of two of the studied samples: sample 1 and sample 2. Samples 1–4 were woven with polypropylene strips in plain weave, while samples 5 had a leno structure made of polyfilaments.

Table 10.1 summarizes the measured characteristics of the samples. The macrostructure of samples 1–3 was similar: the thickness of the strips and densities in warp and weft direction changed, but the strips had one and the same width. Therefore, their fabric weight increased slowly. Sample 5, used to reinforce the bags, had the biggest value of fabric weight.

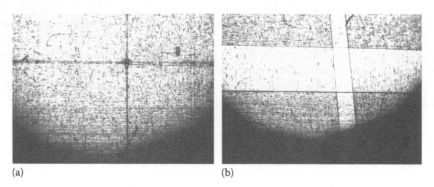

(a)   (b)

FIGURE 10.1

Microscopic pictures (4× enlargement) of the macrostructures of (a) sample 1 and (b) sample 2.

TABLE 10.1

Basic Characteristics of the Samples

Sample	Fabric Weight, g/m²	Warp Density, threads/dm	Weft Density, threads/dm	Strip Width, mm Warp	Weft	Weave	Thickness, mm
1	47	37	30	3	3	Plain	0.07
2	52	37	37	3	3	Plain	0.08
3	66	33	33	3	3	Plain	0.19
4	103	106	37	1.3	1.2	Plain	0.41
5	151	21	21	–	2	Leno	0.53

10.3 Experimental Assessment of the Porosity

The results from the measurements of the pore area, following the methodology described in Chapter 7, are shown in Table 10.2. The mean value is presented together with the variation coefficient and the relative standard error, calculated on the basis of the statistical analysis of the single measurements. Twenty measurements were done for each sample, by contrast with the measurements in Chapter 7, and the choice of a small number of measurements was preconditioned by the exact geometry of the pores of the macrostructures.

Since the measurement of air permeability of sample 5 showed that the macrostructure had no resistance to the air flow (due to the high porosity), the size of the pores of this sample was determined, but only the air permeability of samples 1–4 was evaluated.

Even the mere manipulation of the samples and their observation showed that the macrostructures of samples 1–3 were very unstable with respect to

TABLE 10.2

Pore Area of the Macrostructure

Sample	Mean Pore Area, mm²	Variation Coefficient, %	Relative Standard Error, %
1	0.000273	141.6	42.1
2	0.000189	172.6	51.3
3	0.000103	126.5	37.6
4	0.004254	72.7	21.6
5	11.91	509.2	238.3

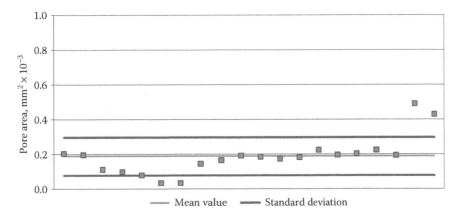

FIGURE 10.2
Pore area of sample 2—single measurements, mean value, and standard deviation.

the displacement of the warp and weft threads, resulting in a substantial difference in the measured pore area even between the adjacent strips. Only the macrostructure of sample 4 was different, and even within the performed small number of measurements, the relative standard error was two times smaller than for samples 1–3.

Figures 10.1 and 10.2 show exemplary results of the single measurements of the pore area for samples 2 and 3. The average values and the standard deviation for each sample are also presented in the charts. Like the results in Chapter 7, the single values of the measured pore area were unevenly distributed and spread beyond the standard deviation. The reason can be found in the non-uniform cross-sectional area of the strips, as well as in the displacement of the strips in the course of the manipulations. The pore area measurement, however, was much faster and accurate due the correct geometric shape of the pores and the absence of protruding fibers.

For one of the samples, the number of single measurements was increased from 20 to 100 to evaluate the effect of the number of measurements on the variation coefficient and the relative standard error. Sample 3 was selected because of its highest fabric weight and thickness, which allowed longer manipulation

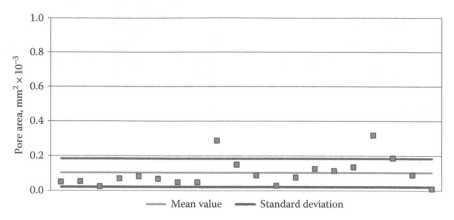

FIGURE 10.3

Pore area of sample 3—single measurements, mean value, and standard deviation.

TABLE 10.3

Pore Area of Sample 3

Number of Single Measurements	Mean Pore Area, mm²	Variation Coefficient, %	Relative Standard Error, %
20	0.000103	126.5	37.6
100	0.000127	123.7	16.0

without displacement of the strips. The expectation was that the number of measurements would not lead to significant changes in the statistical values for this relatively uncluttered macrostructure (strips, plain weave) (Figure 10.3).

The comparison of the results in Table 10.3 showed a difference, however. First, the mean value of the pore area changed, although the difference was after the fourth significant digit. The explanation of this result may be sought not only in the increased number of measurements, but in the greater effect of the manipulation of the specimen under the microscope at five times longer testing. The low friction coefficient between the polypropylene (PP) strips leaded to ease of displacement and therefore to an increment of the distances between the warp and weft threads. That leaded to an augmentation of the average pore size, while the variation coefficient did not change significantly. The absence of a substantial change in the non-uniformity of the pore size, estimated by the coefficient of variation, was an indicative result. Certainly, the relative standard error decreased, which was associated with the increased number of measurements.

The statistical calculations of the results showed, however, that there was no statistical difference between the mean values of the pore area, obtained as an average of 20 and 100 measurements (level of significance $r = 0.05$, ascertained equal variances). That result made the increment of the number of measurements for the rest of the PPWB samples needless.

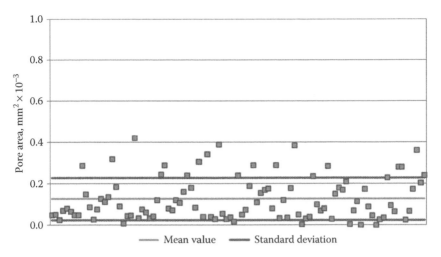

FIGURE 10.4
Pore area of sample 3—100 single measurements, mean value, and standard deviation.

Figure 10.4 shows the results from measurement of the pore area of sample 3: 100 single measurements. The mean value and the standard deviation correspond to data from Table 10.3.

Apart from individual measurement under a microscope, the size of the pore can be determined by calculation, according to Equation 7.1. The results of experimental and theoretical determination of the pore area are shown in Figure 10.5.

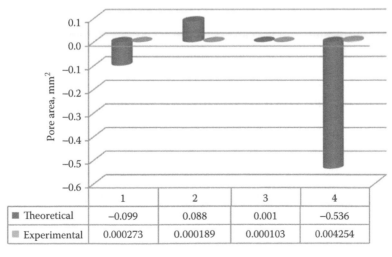

	1	2	3	4
■ Theoretical	−0.099	0.088	0.001	−0.536
▨ Experimental	0.000273	0.000189	0.000103	0.004254

FIGURE 10.5
Pore area of PPWB—comparison between theoretical and experimental results.

The obvious problem with getting negative theoretical values, when applying Equation 7.1, can be explained with the very tight macrostructures of the samples. The fabric is stretched during production, but after that the relaxation process leads to a contraction in both directions.

To use a uniform scale of the graph, the result for sample 5 is not shown, but the calculated pore area for it is 11.77 mm², and the experimentally determined pore area is 11.91 mm².

10.4 Experimental Assessment of the Air Permeability of a Single Layer

It was commented in Chapter 3 that the air flow moves through the pores of the macrostructure only, solely if the woven macrostructure is composed of monofilaments. The polypropylene strips, used for the manufacturing of samples 1–4, have the same effect on the transmission of fluids through the macrostructure, as the monofilaments. Therefore, it was expected that the measured air permeability of the PPWB samples was a function of the pores between the strips only (Luo et al. 2001; Delerue et al. 2003; Gooijer et al. 2003a, 2003b).

The air permeability of the samples in this chapter was measured according to the methodology described in Chapter 5 and represented by the of air permeability coefficient B_p. Four different cases were investigated:

- Air permeability of a single layer
- Air permeability of two parallel layers
- Air permeability of two layers, placed at 90°
- Air permeability of two layers, angled at 45°

The results from the measurement of the air permeability and the respective statistical values are shown in Table 10.4. The student's criterion showed that a statistically proven difference existed (significance level $r = 0.05$) between

TABLE 10.4

Air Permeability of the Single Layer

Sample	Air Permeability Coefficient B_p, m/s	Variation Coefficient, %	Relative Standard Error, %
1	0.0818	29.1	8.3
2	0.0313	41.1	11.7
3	0.1010	20.0	5.7
4	0.4619	7.2	2.0

the averaged values of the air permeability coefficient B_p for the tested macrostructures. The results for B_p were presented again in a function of the characteristics of the macrostructure, like it was done in Chapters 7 through 9.

The fabric cover factor and the fabric areal porosity were not calculated with the standard formulas, presented in Chapter 5, as the linear density of the yarns was not an adequate structural parameter for the polypropylene strips. For that reason, and also because of the discussed process of air permeability through the pores of the macrostructure only, the pore area was used as a characteristic that determined the air flow through the samples. The experimental values of the pore area, measured in this chapter, were used.

The fabric areal porosity was determined by the number m of the pores in the macrostructure (Angelova 2011):

$$V_s = \frac{m.S_p}{S_f}, \%$$ (10.1)

where:
S_f is the area of the sample, m²
S_p is the pore area

The number of the pores in 1 m² of the PPWB samples was calculated in accordance with

$$m = (10P_{wa} - 1)(10P_{wf} - 1)$$ (10.2)

where:
P_{wa} and P_{wf} are the warp and weft densities, respectively, threads/dm

The results from calculation of the fabric areal porosity V_s of samples 1–4 are presented in Table 10.5.

Figures 10.6 and 10.7 show the values of the measured air permeability as a function of the fabric densities in a warp and weft directions, respectively. The reduction in the air permeability with increase in the warp and weft densities was an expected result. The derived linear regression equations with the corresponding correlation coefficient are shown in the figures.

TABLE 10.5

Pore Area of the Macrostructure

Sample	Pore Area, mm²	Number of Pores in 1 m²	Areal Porosity V_s, %
1	0.000273	110331	0.0301
2	0.000189	136161	0.0257
3	0.000103	108241	0.0111
4	0.004254	390771	1.6623

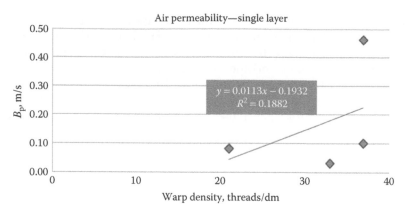

FIGURE 10.6
Air permeability of the PPWB as a function of the warp density.

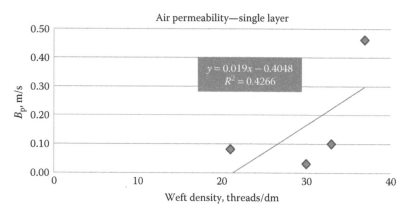

FIGURE 10.7
Air permeability of the PPWB as a function of the weft density.

The inability to establish a statistical correlation between the densities of the macrostructure and its air permeability with a high enough correlation coefficient is evident. The statistical calculations with the linear correlation coefficient indicated that both warp and weft densities and the air permeability coefficient were independent variables.

The increase in the thickness of the macrostructure led to increment of the air permeability of the PPWB (Figure 10.8). The derived linear regression equation had a very high correlation coefficient ($R^2 = 0.93$):

$$B_p = 1.2158\delta - 0.00599 \tag{10.3}$$

This result differed from the results in Chapters 7 and 8, where the correlation coefficient for the same dependence was $R^2 = 0.5$ (in Chapter 8—for the reduced number of six samples).

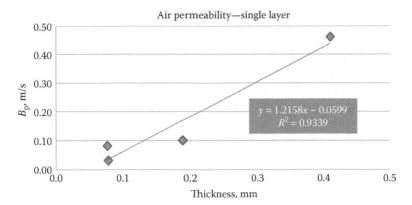

FIGURE 10.8
Air permeability of the PPWB as a function of the thickness.

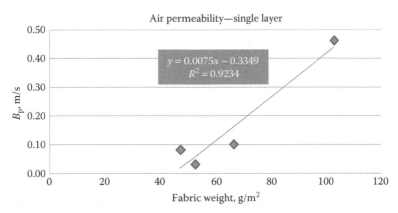

FIGURE 10.9
Air permeability of the PPWB as a function of the fabric weight.

The same result was obtained for the effect of the fabric weight on the coefficient of air permeability of PPWB (Figure 10.9).

In Chapter 7, the fabric weight was analyzed as a more complex characteristic of the macrostructure than the thickness as regards the transfer of fluids. Here, the correlation coefficient of the derived regression equation was again very high ($R^2 = 0.92$). It is of the same order as in the established linear relationship between the fabric weight and the air permeability in Chapters 7 ($R^2 = 0.89$) and 8 ($R^2 = 0.79$, for six samples). The derived regression equation is

$$B_p = 0.075 m_s - 0.3349 \tag{10.4}$$

The investigation of the specific macrostructure, woven with polypropylene strips, allowed an analysis of the effect of the size of the pores (spaces) between the threads on the air permeability to be made (Figure 10.10). In all other cases, such an analysis would not be correct because of the possibility of the air to flow through the pores of the mesostructure. The derived regression equation had a very high coefficient of correlation ($R^2 = 0.97$). There was a logical tendency the transferred flow through pores with a larger area to be with higher flow rate.

This trend was confirmed by the graph in Figure 10.11, which presents the functional relationship between the air permeability coefficient and the areal porosity of the samples. The derived linear regression equation had a very high coefficient of correlation ($R^2 = 0.97$) as well.

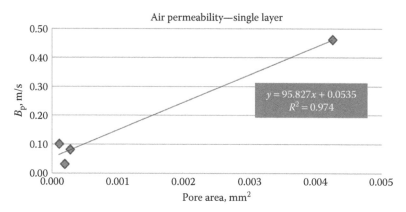

FIGURE 10.10
Air permeability of the PPWB as a function of the pore area.

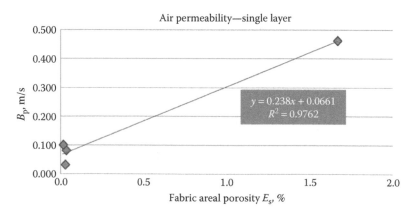

FIGURE 10.11
Air permeability of the PPWB as a function of the fabric areal porosity.

However, the graphs in Figures 10.10 and 10.11 does not allow accurate assessment of the impact of the studied parameters on B_p, due to the very close values of the pore size and areal porosity, respectively, of three of the four samples. Although the correlation coefficient is high, more samples should be examined so that intermediate values for pore size and areal porosity to confirm the presence of the derived functional dependence with high R^2. In any case, the statistical test of the linear correlation coefficient with the student's criterion confirmed the existence of a statistically proved relation between the investigated characteristics in Figures 10.10 and 10.11.

The performed study on the air permeability of PPWB allowed one more analysis to be done: on the effect of the air transfer through the mesostructure. The air permeability coefficients for macrostructures with one and the same areal porosity (or fabric cover factor) were compared: B_p of sample 2 from the upholstery textiles (in Chapter 9) was compared with B_p of sample 1 from the set of PPWB, as both samples had a real porosity of 0.03%. Similarly, the air permeability of sample 3 of the upholstery textiles was compared with sample 2 of PPWB, as both macrostructures had 0.02% area porosity. The result of the comparison of the air permeability coefficients of the samples is shown in Figure 10.12.

It is noteworthy that the air permeability coefficient of the upholstery textiles was higher in comparison with the polypropylene macrostructures for packaging. Having the same areal porosity (or fabric cover factor), the obvious reason was the passage of air through the mesostructure of the upholstery textiles and the impossibility of such passage in the samples for packaging.

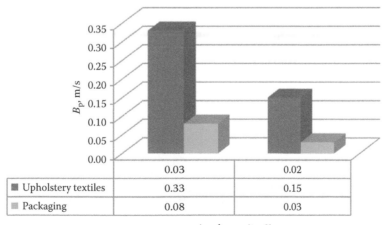

	0.03	0.02
■ Upholstery textiles	0.33	0.15
■ Packaging	0.08	0.03

Areal porosity, %

FIGURE 10.12
Comparison between B_p of macrostructures with equal areal porosity and permeable and non-permeable mesostructure.

10.5 Experimental Assessment of the Air Permeability of Double-Layered Systems

In this part of the study, the air permeability of systems of two identical macrostructures of PPWB was experimentally determined.

A polypropylene mesh was inserted between the two layers so as to eliminate the close contact between the samples during testing. The mesh had no resistance to the passing air flow and was described in Chapter 8. The surface of the samples was 10 cm²; the difference in pressure on either side of the system was set to 100 Pa. Fifty single measurements were made for each system.

Student's criterion was used to evaluate the difference between the mean values of the samples' air permeability within each group. It was found that there was no statistically proven difference only between B_p of samples 1 and 3 in the systems of two layers, angled 90°. For all other cases, the mean values of the air permeability coefficient were different.

The results are displayed in Figure 10.13, where the measurements of the single layer's air permeability are also visualized.

It is logical to expect that any overlapping of two textile layers will block part of the pores of the macrostructures. Therefore—as it can be seen in Figure 10.13—the air permeability of a system of two layers will always be less than the air permeability of the single layer.

The comparison of the results for the transmission of air through a system of two layers and two layers at an angle of 90° indicates the absence of statistical evidence for a difference between the average values of B_p for samples 1–3. This result may be explained by the particularity of the samples mesostructure: the width of the strips, used for warp and weft threads, was the same (3 mm). The warp and weft densities of samples 2 and 3 were also equal, and the difference was small for sample 1. Therefore, the rotation of one layer

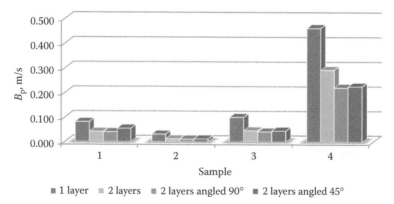

FIGURE 10.13
Air permeability of the PPWB samples.

TABLE 10.6

Air Permeability of Systems of Two Identical Layers

Sample	Air Permeability Coefficient B_p, m/s		
	Two Layers	Two Layers, Angled 90°	Two Layers, Angled 45°
1	0.0419	0.0409	0.0551
2	0.0123	0.0117	0.0125
3	0.0461	0.0419	0.0453
4	0.2932	0.2211	0.2257

at 90° was not important for the studied permeability. A proven statistical difference was established for sample 4, where both substantial differences between the warp and weft densities and the width of the strips in both directions existed (Table 10.6).

Further statistical calculations were performed for comparison of the mean values for the air permeability of system of layers, angled 90° and 45°. The application of student's criterion showed that there was an evidence for statistical difference between the values for samples 1 and 4, while the rotation of the layers at different angles for samples 2 and 3 did not affect B_p. This effect was perhaps based on the characteristics of the macrostructure of samples 1 and 4—the warp density was different from the weft density, unlike the other two samples, which were balanced out by densities. Thus, placing the two layers at an angle of 45° resulted in a less overlapping of the pores and greater permeability as compared to the samples with equal densities in both directions.

10.6 Theoretical Assessment of the Air Permeability of Double-Layered Systems

In this part of the study on PPWB, theoretical and experimental results, obtained by measuring the coefficient of air permeability B_p of systems of two identical layers, were compared. The following theoretical models were used for the calculation of the theoretical values:

- The model of Clayton (1935) described and applied in Chapter 8
- The developed theoretical model in Chapter 8

The results of the comparison between the theoretical and experimental values of the air permeability coefficient for a system of two identical layers are shown in Figure 10.14. Figures 10.15 and 10.16 summarize the results for systems of two identical layers, angled 90° and 45°, respectively. The theoretical results in Figures 10.14 through 10.16 are equal, as the model of Clayton

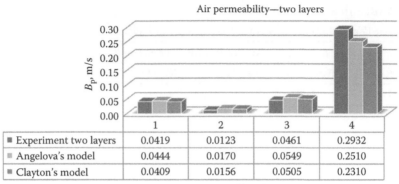

FIGURE 10.14
Air permeability of systems of two identical layers of PPWB.

	1	2	3	4
■ Experiment two layers	0.0419	0.0123	0.0461	0.2932
■ Angelova's model	0.0444	0.0170	0.0549	0.2510
■ Clayton's model	0.0409	0.0156	0.0505	0.2310

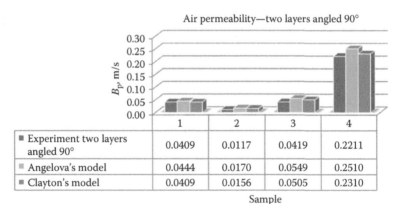

FIGURE 10.15
Air permeability of systems of two identical layers of PPWB, angled 90°.

	1	2	3	4
■ Experiment two layers angled 90°	0.0409	0.0117	0.0419	0.2211
■ Angelova's model	0.0444	0.0170	0.0549	0.2510
■ Clayton's model	0.0409	0.0156	0.0505	0.2310

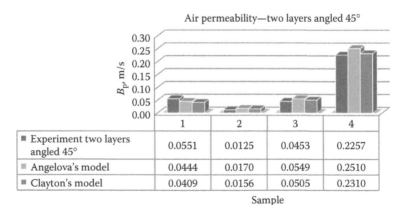

FIGURE 10.16
Air permeability of systems of two identical layers of PPWB, angled 45°.

	1	2	3	4
■ Experiment two layers angled 45°	0.0551	0.0125	0.0453	0.2257
■ Angelova's model	0.0444	0.0170	0.0549	0.2510
■ Clayton's model	0.0409	0.0156	0.0505	0.2310

and the derived model in Chapter 8 (quoted on the chart as *Angelova's model*) reflect only the number of layers and the identity of their macrostructure and mesostructure, but not the geometric position one to the other. It is therefore interesting to evaluate to what extent the models are close to the experimentally determined coefficient of air permeability of the system of two identical layers at different position of the layers.

The calculation of the relative error between the results of Angelova's model and the experimental data showed that the deviation ranged from 5.6% for sample 1 to 27.6% for sample 2, and for samples 3 and 4 the relative error was 16% and 16.8%, respectively. The deviations are not very high and it can be concluded that the developed theoretical model can be used for practical engineering calculations.

The analysis of the results showed that the two theoretical models applied leaded to deviations from the experiment, while the values of Clayton's model were a little closer in most cases to the experimental results, compared to Angelova's model. Generally, the two models gave a bit higher results than the measured values of B_p. A possible explanation of this result should be the absence of air flow through the pores of the mesostructure of the polypropylene samples, which is not reflected by the theoretical models.

10.7 Summary

A systematic study of the air permeability of systems of layers of woven macrostructures, which mesostructure did not allow the transfer of air, was performed. Results on the effect of the pores of the macrostructure (experimentally determined) on the air permeability of the macrostructure were obtained.

A statistical relationship between the air permeability coefficient B_p and the areal porosity, determined by measuring the area of the pores and the number of pores per unit area of the macrostructure (again in the absence of transfer of air through the mesostructure), was derived.

The results on the effect of structural, geometric, and mass characteristics of the woven macrostructure on the transfer of air through a single woven layer in Chapters 7 through 9 were complemented and enriched with a range of woven macrostructures with different applications.

Original results on the effect of the air transfer through the woven mesostructure were obtained on the basis of comparison between the air permeability coefficient of upholstery textiles and textiles for packaging with equal areal porosity.

Original results on the effect of the position of two identical layers on the air permeability coefficient were obtained: for coincidence of the direction of warp and weft threads, for two layers, angled 90°, and for two layers, angled 45°.

Verification of the developed theoretical model in Chapter 8 for the air permeability of a system of textile layers was carried out. The verification of the model with a new set of woven macrostructures showed that the relative error of the theoretical results, compared to experimental ones, was in the range of 5%–16% for 75% of the cases studied.

11

Experimental Investigation of Textile Macrostructures in the Indoor Environment

11.1 Introduction

The comfort, health, and productivity of people indoors depend to a great extent on the indoor environmental parameters. Moreover, it is necessary for the indoor environment to have stable parameters, as it concerns temperature values and relative humidity (RH).

The investigation of Fang et al. (2000) concluded that the fluctuations in the RH are as unpleasant for the inhabitants as the temperature changes. Frequent or high fluctuations in the RH or too high RH can provoke damage to the building construction and furniture: structure deformations (shrinking or dilatation), mould growth, or cracks appearance.

People assess the indoor environment with too high or too low RH as unpleasant one. The reasons are related with the appearance of nasty odors, condensation on surfaces (including on human skin), or static electricity. The rapid growth of the bacteria due to high RH may provoke health problems as well (Bornehag et al. 2003; Naydenov et al. 2006).

As it has been already discussed in Section I of this book, the textiles affect all aspects and parameters of the indoor environment:

- *Thermal environment*: The textiles act as insulating layers in the case of floor coverings, upholstery, and so on. In the case of coverings, they create a microenvironment around the human body, whose temperature, in general, is higher than the room temperature.

- *Atmospheric environment*: All textiles in the indoor environment, including the clothing of the inhabitants, participate in the exchange of moisture in the air.

- *Acoustic environment*: The textiles reduce the noise in the indoor environment.

- *Visual environment*: The textiles create aesthetic spaces and give them personality.

Thermophysiological comfort is associated with the thermal environment as an aspect of the indoor environment. Textiles and clothing, however, affect the atmospheric environment as well through their effect on the RH. On the other hand, the moisture content of the micro-, meso-, and macrostructure of the textiles influences the behavior of textiles and clothing as regards the transfer of air and heat.

Despite their function in the indoor environment as direct or indirect insulation layers (see Chapter 2), the textiles can play the role of moisture buffering materials indoors. Moisture buffering is the ability of the surface materials to reduce the fluctuations in the room air humidity through absorption and desorption of moisture. Actually, the inside construction of the buildings and the materials used for finishing and decoration of walls, ceilings, floors, and stairs, together with the furniture and all other furnishing, have influence on the RH in the indoor environment. On the other side, RH of the indoor environment influences the moisture content of the materials and their abilities to act as moisture buffers in a particular environment (temperature, ventilation technique used, etc.)

The moisture buffering role of the textile macrostructures in the indoor environment is associated with the partial pressure. When the partial pressure of water vapor in the microstructure (i.e., in cotton fibers) becomes equal to the partial pressure of water vapor in the air, the processes of absorption or desorption cannot occur. The presence of the pressure gradient, however, causes the transmission of water vapor from the material to the environment or vice versa.

The ability of the materials to influence the RH of the indoor environment is particularly important for the human health, especially in the case of bedrooms. The moisture buffering effect of the textiles plays a significant role for the indoor climate and survival of house dust mites, a common cause of allergies and asthma among children and adults. Moisture buffering has its importance for the better understanding of the risks for mold growth on walls and other type of biological growth in indoor surfaces, which is an additional cause for diseases among the inhabitants. It is also suspected that high moisture content of the surface materials in rooms increases the risk of chemical emissions and volatile organic compounds (VOCs) from the indoor surfaces to the air (Bornehag et al. 2003; Naydenov et al. 2006). Apart from all these health aspects, moisture buffering is also important to be taken into account when designing air conditioning systems, calculating energy demands, or simulating buildings performance (Fang et al. 2000; Naydenov et al. 2008).

There are different types of materials that are exposed to the indoor air. They are completely heterogeneous by their nature: from the lightweight textile curtains to the heavy concrete in the building construction. Greater part of the materials used indoors consists of polymers (i.e., cellulose) like wood, cotton, and paper. However, the materials exposed to the indoor air often have, in one way or another, a surface coating (Rode et al. 2002; Svennberg et al. 2004).

The literature review has shown that there is a lack of findings concerning the quantities and distribution of the exposed areas of surface materials in real and particular indoor environments. Moreover, the correct evaluation of the influence of different moisture buffering materials on a certain indoor environment requires assessment of the indoor environment parameters in different zones of the room.

The new possibilities for using modern software packages for design of heating, ventilation, and air conditioning (HVAC) systems and building simulation require an understanding of the influence of the indoor materials on the air RH in the rooms. More practical is the need to introduce in the calculations the properties of all materials in the indoor environment, including the textiles. Several studies (Harderup 1999; Rode et al. 2002; Svennberg et al. 2004) have shown that there is a need of better understanding and deeper knowledge about the properties of textile materials and their influence on the RH of the indoor air.

11.2 Materials in Contact with the Air in the Indoor Environment

The RH represents the ratio between the amount of water vapor in the air and the saturation value (maximal amount). The saturation point of the air is as high as higher is the temperature of the air.

RH is often defined by the water vapor pressure:

$$RH = 100 \frac{p_v}{p_{v,sat}} \qquad (11.1)$$

where:

p_v, Pa, is the ambient vapor pressure
$p_{v,sat}$, Pa, is the saturation vapor pressure

The ideal gas low gives the relationship between the ambient vapor pressure p_v and the vapor concentration ρ_v, kg/m^3:

$$p_v = \rho_v RT \qquad (11.2)$$

where:

T is the temperature, K
R is the gas constant of water vapor (462 J/kgK)

The saturation vapor pressure can be expressed by Equation 11.3:

$$p_{v,sat} = 610.5 . \exp \frac{17.269\theta}{273.3 + \theta} \qquad (11.3)$$

where:

θ is the temperature, °C

The vapor content of the indoor air changes over time. If the temperature dependence of the volume expansion of the air is neglected, the changes of the vapor content of the indoor air can be described analytically by Equation 11.4 (Plathner and Woloszyn 2002):

$$\frac{\partial v_i}{\partial t} = \frac{\sum G_{ms} - \sum_{j=1}^{n} g_{mbj} A_{sj} - q(v_i - v_e) - \beta_v(v_i - v_{ss})A_c}{V} \tag{11.4}$$

where:

G_{ms} is the moisture supply, kg/s

g_{mb} is the adsorbed or desorbed moisture flux of the surface material j, kg/(m²s)

A_{sj} is the area of the surface material j, m²

q is the ventilation flow rate, m³/s

v_i is the vapor content of the indoor environment, kg/m³

v_e is the vapor content of the outdoor environment, kg/m³

β_v is the surface moisture transfer coefficient, m/s

v_{ss} is the saturation vapor content at the surface, kg/m³

A_c is the area of condensation, m²

V is the volume of the room, m³

The moisture supply G_{ms} is composed of two parts:

- *Constant component (or relatively constant)*: It is related with the moisture supply from the inhabitants (including plants and pets).
- *Pulse component*: It is formed by short moisture supply pulses, which arise from different indoor activities like baths, showers, cooking, laundry, and ironing. These activities are spread over short periods of time during the day.

The moisture absorption or desorption in the different surface materials at a given time g_{mb} reflects their ability to buffer the variations in the RH of the indoor air (Svennberg et al. 2004). All indoor materials—building envelope, furniture, interior textiles, and other furnishing—are described by this term. In most cases, the moisture buffering ability of a hygroscopic material in real conditions will be lower than the basic material properties suggest. This is due to sorption isotherm hysteresis and the fact that in most real situations, the surface materials of the indoor environment have a surface coating or treatments that reduce their moisture capacity (Rode et al. 2002).

The flow rate q of the air, supplied by the ventilation systems, guarantees sufficient amount of fresh air for the building and its inhabitants and is responsible for removing emissions (from activities, people, and materials), excess moisture, smells, and hazardous substances. The impact of the HVAC systems on the RH of the room depends on the ventilation flow rate and the vapor content of both indoor and outdoor air (Stankov 2003). When the air exchange rate is high, it preconditions the moisture content of the indoor air to a high degree (Plathner and Woloszyn 2002). Therefore, the impact of moisture buffering abilities of the surface materials indoors is larger when the air exchange rate is low.

Surface materials in the indoor environment are all materials, whose surface area is able to play the role of sorption surface or to participate in sorption processes (Svennberg et al. 2004). Their sorption capacity is usually presented as a function of the RH in the room. Rode et al. (2002) determined the cellulose-based materials as those with the greatest ability to act as moisture buffers indoors, followed by the cellular concrete for RH over 80%. However, Rode et al. (2002) did not comment on the abilities of textiles to participate in the sorption processes in the indoor environment.

11.2.1 Wood and Wood-Based Materials

In the indoor environment, wood and wood-based materials are either part of the furniture or part of the building construction. Depending on the type of the architecture, it could dominate over other building materials (i.e., in wooden houses). The wooden surfaces are usually coated, though other option exists as well (i.e., non-coated wooden ceiling).

The wood fibrous composition preconditions its heterogeneous structure. The hygroscopic abilities depend to a high degree on the orientation of the wood fiber strands compared to the moisture source. Rode et al. (2002) commented that the moisture uptake rate is 20 times higher in parallel, than in tangential direction to the fibers. Besides, the authors mentioned that softwood (coniferous tree) has often higher moisture permeability than hardwood (deciduous tree).

Surfaces of wood and its derivatives have a good capacity to compensate moisture. At the same time, moisture in the wood is one of the main reasons for the destruction of the material: it causes biological growth of organic substances, and thus, the wooden surface may become a source of VOCs or MOCs (see Chapter 2).

11.2.2 Textiles

Sorption capacity of textiles in the indoor environment is highly dependent on the properties of the microstructure (i.e., the used fibrous materials), as well as the meso- and macrostructure. The assessment of the joint behavior of layers of fabrics (i.e., in mattresses) or of combination between textile and other materials (i.e., with wood or plastics in the furniture) is quite complex.

In general, the natural fibers have a large ability to compensate the moisture in the air, especially for short periods of time. At the same time, the interior textiles (upholstery textiles, carpets, and blankets) are the main habitat for dust mites, which are among the main causes of allergies and asthma (Bornehag et al. 2003; Naydenov et al. 2008).

11.2.3 Paper and Paper-Based Materials

Generally, three types of paper and paper-related materials can be found in the indoor environment: bound paper (books, magazines, newspapers), loose paper (sheets of papers, usually on desks), and wallpaper. Bound paper forms one major part of the paper in the indoor environment. Paper wallpapers are frequently big area of surface materials indoors, but they are usually coated to be hydrophobic.

Svennberg et al. (2004) has shown that storage of paper in the indoor environment is very important on account of its moisture-buffering abilities. The authors found that the vapor content behind bookshelves in an office room was lower than in the center of the room. Moreover, the type of the shelves (books storage) also influenced the capacity of paper-based materials to participate in sorption and desorption processes in the room.

11.2.4 Gypsum and Gypsum Boards

Gypsum boards are suitable for finishing the interior walls and ceilings of the building construction. They are increasingly used in buildings, especially in the Nordic countries, where the wet indoor finishing processes are avoided. Usually, a core of gypsum is covered with thick paper on both sides and afterward wallpaper or paint is applied.

Gypsum plasters are typical for building structures in the southern countries. Like gypsum boards, they play an important role in buffering the moisture indoors. For a short period of time, however, they themselves are a source of moisture in the building.

11.2.5 Concrete and Ceramics

The concrete (including aerated concreted) and ceramic materials (ceramic tiles) have very big relative surface area, compared to other materials, but almost zero effect as indoor air moisture buffers. The reason is that they are usually covered, coated, or glazed and have no direct contact with the room air.

11.2.6 Other Materials

Surface coatings, such as paint, wax, and plastic films, are often based on polymers. Due to such finishing materials, the hydrophilic abilities of wood, concrete, and paper can be significantly reduced (Plathner and Woloszyn 2002; Harderup 2005).

Other materials, used indoors, like metals and glass, do not add to the moisture buffering capacity of the building and furnishing and can be neglected. Plastic materials, which can be frequently find indoors in floorings and furniture, have no homogeneous moisture properties and in many cases can be also neglected as moisture buffers (Svennberg et al. 2004).

11.3 Distribution of Moisture Buffering Materials in Real Enclosures: Offices and Hotel Rooms

The main aim in this part of the study was to investigate the behavior of textile materials in the indoor environment in the aspect of their interaction with water vapor in the air. A statistical inventory of the materials that can act as moisture buffers in four types of enclosures was made. Two offices at the Technical University of Sofia and two hotel rooms on the Black Sea coast were selected. The choice of the buildings was determined by (Angelova 2009) the following:

- Differences in their use
- Differences in the external climatic conditions
- Differences in the period of occupation

The region of the Black Sea coast is characterized by high humidity and the presence of constant winds. Sofia region is characterized by moderate atmospheric humidity and periodic winds. The office rooms are inhabited year round, unlike hotel rooms, which are subject to seasonal use (in summer). This has determined the difference in the architecture of the buildings.

The common thing between the selected buildings was that the building constructions include concrete for ceiling and flooring, and bricks for the walls. The walls and ceilings were finished with gypsum and paint coated. All investigated rooms had orientation toward south.

The two office rooms had one and the same area, but office 1 was occupied by one person, while office 2 was occupied by two persons, which led to differences in the room furniture and finishing. Metal Persian blinds were used on the windows in the two offices. The floor in office 1 was covered by laminated parquet, while the floor in office 2 was covered by ceramic tiles.

The two hotel rooms had one and the same area as well, but room 1 was for families with one child (one double bed and one single bed), while room 2 had one double bed only. Laminated parquet was used as a floor covering. The windows were covered with two layers of curtains of two different types of thickness and porosity.

Two types of wooden materials were found in the indoor environment: floor covering and furniture. All wooden surfaces were coated. Untreated wood was not found in the four investigated enclosures.

The textiles were estimated as two types as well: textiles, being in a *direct* contact with the inhabitants (bedding and upholstery) on the one hand, and textiles, which were *indirect* insulation layer (curtains and carpets), on the other.

The reason for this separation is that carpets and curtains are, in general, single layer fabrics, while bedding and upholstery textiles are mainly complex systems of textile materials. For example, a mattress in the hotel room had as a first layer quilted cotton fabric for best feeling. The second layer was thick cotton felt, which absorbs the body heat. The third layer was coconut felt for natural comfort and stability over the last layer of the springs. Among the springs, large volume of air existed. It was a really hard task to estimate the moisture ability of the system of these materials, though all of them (except the metal springs) could absorb moisture from the air.

Paper and paper-based materials were also divided into two groups: *bound* paper (books and magazines) and *loose* paper (newspapers and sheets, which could be frequently found in offices). The area of the loose paper was estimated with the assumption that it was placed in stacks, following the method described in Rode et al. (2002): the base area was equivalent of an A4 paper and the height of 0.1 m. According to this assumption, each 1 m of the stack was equivalent to an exposed area of 1.6 m^2.

Table 11.1 shows the results from the inventory. All areas of moisture buffering materials, exposed to the indoor air in the rooms, were measured.

The distribution of the moisture buffering materials in the office rooms is presented in Figures 11.1 and 11.2. The analysis of the results showed that the largest areas in the offices belonged to wood and wood-based materials as well as on the gypsum walls and selling. However, the wooden furniture and the laminated parquet cannot be estimated as moisture buffering materials for the particular case, as they were coated. The difference of 25% more wood and wooden materials in office 1 was due to large areas of enclosed cupboards, including wall cupboards, which were used for books and paper storage. The last explained was the appearance of 17% more books and paper in office 2, as they were mainly stored in open shelves.

If the influence of the gypsum on the walls and ceiling of the two offices on the moisture content of the indoor air was neglected, the paper and paper-based materials were the main moisture buffers for short-term periods in the investigated offices. It has to be mentioned that Scandinavian offices are distinguished by having non-coated wooden furniture (Rode et al. 2002), and it is additional big source of moisture buffering material.

TABLE 11.1

Materials in the Selected Offices and Rooms

Materials	Wood and Wood-Based Materials—Area, m²		Textiles—Area, m²		Paper and Paper-Based Materials—Area, m²		Gypsum and Gypsum Boards—Area, m²	Concrete and Ceramic Materials—Area, m²	Windows—Area, m²
Enclosures	Floor	Furniture	Carpets and Curtains	Bedding and Upholstery	Bound Paper	Loose Paper			
Office 1	15	44.9	0	4.3	9.8	2.1	50.1	0	5.1
Office 2	0	28.4	1.2	0.8	33.2	1.3	50.1	15	5.1
Room 1	16	22.47	30.2	6.2	0.3	0.3	50.5	0	8
Room 2	16	20.7	60.4	4.6	0.2	0	41.5	0	16

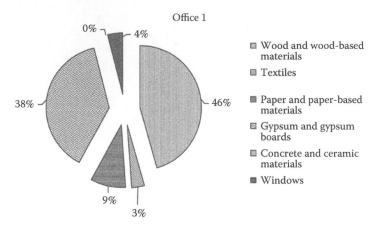

FIGURE 11.1
Distribution of materials, exposed to indoor air—office 1.

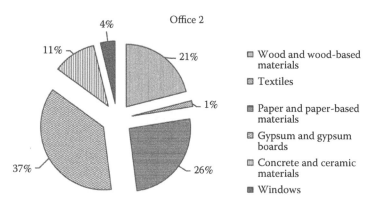

FIGURE 11.2
Distribution of materials, exposed to indoor air—office 2.

The analysis of the inventory of the indoor materials in the two hotel rooms showed that a new group dominated over the paper materials—the textiles. While the percentage of textiles in the offices was 1%–3%, Figures 11.3 and 11.4 showed that the area of the textiles increased to 25% for room 1 and 41% for room 2. The increased proportion of the textiles in room 2 was due to the larger area of windows: great part of an outer wall was replaced by floor-to-ceiling panoramic window.

The same conclusions, drawn for the influence of the gypsum and wooden surfaces, could be taken here as well. Wooden surfaces of furniture and parquet were practically impermeable due to the coatings. Gypsum on the walls and ceilings could hardly influence the RH of the indoor air in short-term

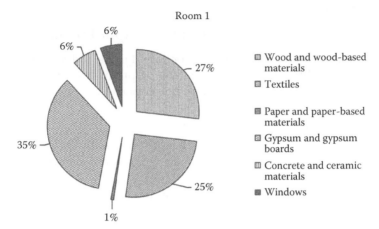

FIGURE 11.3
Distribution of materials, exposed to indoor air—room 1.

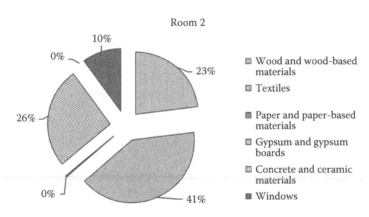

FIGURE 11.4
Distribution of materials, exposed to indoor air—room 2.

periods. Therefore, textiles were the main factor responsible for the moisture buffering in the rooms.

In conclusion, the performed measurements in four rooms of different type have shown that different materials, which may affect the air humidity, can be found in the indoor environment. The statistical analysis has indicated that besides building construction materials (mainly gypsum), paper and textile surfaces have the largest area in the indoor environment. It was found that the materials, based on paper, dominated the office environment, while textiles had the largest area in the hotel rooms.

11.4 Interaction of Textiles with Water Vapor Indoors

The textile fabrics and items, which can be found in the indoor environment, have different abilities to participate in the sorption/desorption processes. This ability is determined by the following properties:

- *Of the microstructure*: The fiber type
- *Of the mesostructure*: Type and structure of the threads (i.e., spinning method used), twist level of the spun yarns or twisted polyfilaments
- *Of the macrostructure*: Type of the fabric (woven, knitted, nonwoven), number of fabric layers (single, double, multiple), pattern, finishing, and so on

The presence of nanoparticles or nano-layers has to be also taken into account.

The type of textile material and, to some extent, the mesostructure determine to a great extent the equilibrium moisture content, while the type and properties of the meso- and macrostructure determine the latent heat exchange through the textile. In equilibrium, the moisture content of the textile materials depends on the temperature and the RH of the environment. High temperature or low humidity reduces the moisture content of the textiles. It has a maximal value in an environment of low temperature and high humidity (Angelova 2009).

Two types of water can be found in the microstructure of the textiles (Armour and Cannon 1968):

- *Free water*: It is placed in the lumen of some fibers (mainly cellulose-based fibers) or in the pores between fibers.
- *Bonded water*: It is situated in the pores of the fiber wall or it is chemically bonded to either hydroxyl groups or carboxylic acid groups in fibers.

The moisture content strongly preconditions the type of water in the fibers (Whitaker 1977): at moisture contents below 20%, only the water, bonded to the hydroxyl groups, remains in the fibers. If the moisture content increases to 20%–40%, the pores in the fiber wall start their action as well. This corresponds to temperature of 23°C and RH > 90%. Free water can be observed just at very high moisture content.

Due to its ability to participate in the process of water absorption from the room air, the textile fibers increase their volume, swelling. The reason is that the water molecules penetrate between the hydrogen-bonded fibrils in the fiber wall and the amount of bonded water increases, while the degree of internal bonding of the fiber wall decreases. The process of water desorption provokes the opposite: the quantity of the bonded water decreases, followed by better internal bonding of the wall's fibrils.

The basic properties of the textile materials often depend on the moisture content: a dynamic nonlinear characteristic, which is temperature dependent. The hysteresis effect is a physical phenomenon during sorption: the quantity of adsorbed or absorbed moisture is different than the quantity of the desorbed by the fibers moisture. The result is that at a certain RH, the fibers' moisture content is different in absorption, coming from dry conditions, than in desorption, coming from humid conditions. Depending on the history, the moisture content can be anywhere between the two boundary curves of the hysteresis (Ashpole 1952; Ülkü et al. 1998; Jeremy et al. 2003).

When the focus is particularly on the textiles, the hysteresis effect that appears is related to the hygroscopic properties of the fibers. Thermal energy, called as well *heat of desorption*, is needed to remove the water from the fibers (Whitaker 1977). On the contrary, when the fibers absorb water, *heat of absorption* is released. The result is that, due to the heat of absorption, a difference in water vapor pressure (or RH) appears between absorption and desorption.

There are different theories about the background of the hysteresis effect. The main theory explains that hysteresis arises from the presence of independent microscopic domains (Wadsö et al. 2004). These domains can be in two states: *sorbing* or *nonsorbing*. The switch between the two states requires an energy barrier to be crossed, and it is done during absorption or desorption of water (Svennberg and Wadsö 2008). According to another theory, the *bottleneck theory*, the shape of the microscopic domains actually controls the hysteresis (Carmeliet et al. 2005). The microscopic domains have different sizes and the capillary pressure makes the smaller domains to be more ready to absorb water than larger domains. There are opinions that the hysteresis is due to the hydroxyl groups, which bind the water (Adamson 1990). This theory supposes that the cellulose molecules form groups, which are weakly bonded to each other. Some of the bonded hydroxyl groups become free during the water absorption and they bound more water. In desorption, the opposite occurs (Svennberg and Wadsö 2008).

The last theory uses the swelling stresses during absorption to explain the hysteresis effect. It is based on the irreversible plastic deformations, which appear due to the swelling stresses. These stresses arise in the fiber wall because crystalline cellulose does not swell, while the non-crystalline cellulose regions increase their volume (swell). The deformations remain elastic if small amounts of water are absorbed. At larger amounts of water, however, the swelling stress may exceed a limit point, which causes plastic deformations. Due to this, the weak bond between the groups of cellulose molecules can be broken and additional free hydroxyl groups to absorb water (Carmeliet et al. 2005).

11.5 Sorption Isotherms

The aim of the measurement was to determine the sorption isotherms and the hysteresis ratio of five woven macrostructures used in indoor environment: two of them as curtains and three as upholstery textiles. Table 11.2 presents the main properties of the samples, sorted in accordance with the increment of the fabric weight. All macrostructures were single layered.

The measurement was performed using a climatic cabinet that allowed the control of the RH and the methodology of Anderberg and Wadsö (2004) was followed. The RH was changed from 10% to 95% and back to 10%. The measurement of the weight of the samples (20 mm diameter) continued until equilibrium was reached for the respective RH. Three samples were tested for each macrostructure.

Figures 11.5 and 11.6 show the hysteresis effect for two of the investigated samples: sample 1 and sample 3.

TABLE 11.2

Properties of the Samples Used

Sample	Material Composition	Weave	Application	Thickness, mm	Fabric Weight, g/m²
1	100% cotton	Plain	Curtain	0.40	245
2	55/45% cotton/flax	Plain	Curtain	0.51	264
3	50/50% wool/ viscose	Plain	Upholstery textile	1.08	398
4	80/20% wool/ polyamide	Jacquard weave	Upholstery textile	0.98	410
5	100% wool	Twill 2/2	Upholstery textile	1.0	441

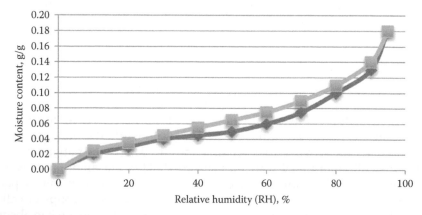

FIGURE 11.5
Sorption isotherms for sample 1 (cotton 100%).

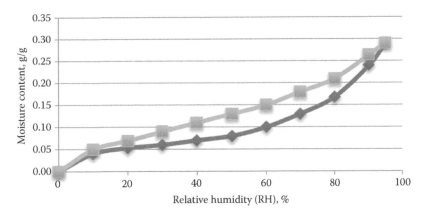

FIGURE 11.6
Sorption isotherms for sample 3 (wool/viscose 50/50%).

The samples containing cotton (samples 1 and 2) showed similar sorption isotherms. As it was expected, the cotton fibers absorbed less moisture from the air, compared with the samples, containing wool fibers (samples 3–5). Besides, the hysteresis effect of the cotton-based textiles was less manifested in comparison with the wool-based macrostructures.

Although under normal climatic conditions (RH = 65 ± 5%, $t = 20 \pm 2°C$) viscose fibers have a smaller capacity for moisture absorption than the wool fibers, sample 3 showed higher values for moisture content (for RH = 95%) (Figure 11.6). The presence of polyamide fibers in sample 4 provoked the opposite effect. The pattern of the three wool-based samples was similar: with the largest hysteresis in the middle of the RH range and a decreasing hysteresis with higher RH, giving the lowest hysteresis effect at 90% RH.

On the basis of the sorption and desorption isotherms, the hysteresis ratio H was calculated according to the formula (Okubayashi et al. 2004):

$$H = \frac{u_{\text{des}} - u_{\text{abs}}}{u_{\text{abs}}} \qquad (11.5)$$

where:

u_{des} and u_{abs} are the respective values for moisture content during desorption and sorption processes for the respective RH

Figure 11.7 demonstrates the result for the calculated hysteresis ratio of the investigated samples.

The analysis showed that all samples demonstrated largest hysteresis effect around the middle of the studied interval of the RH at RH = 40%–70%. This interval corresponds to the levels of RH, maintained in most buildings by the systems for HVAC. The results clearly showed that textiles can be good moisture buffers for this particular interval of RH, by absorbing water vapor from the air at some moment and release it at another.

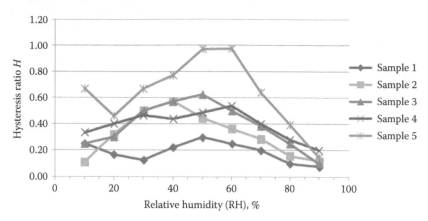

FIGURE 11.7
Hysteresis ratio *H* of the measured samples.

In conclusion, frequent or high fluctuations in the RH or too high RH are not acceptable from the point of view of both inhabitants and possible construction damage of buildings. Textiles are excellent moisture buffers in indoor environment due to their often high moisture capacity and their ability for rapid absorption and desorption of moisture compared with other surface materials exposed to indoor air.

11.6 Summary

Results on the distribution of the moisture buffering material in the indoor environment were obtained. The survey involved four enclosures: two offices and two hotel rooms.

The hysteresis effect of sorption and desorption of water vapor from the air of five types of textiles was determined. It was found that the hysteresis effect was highest in the range of RH = 40%–70%, which is the range maintained by the HVAC systems in buildings.

Both the statistical distribution of textile items in the indoor environment and the study on the hysteresis effect have proven that textiles are effective buffer of the air RH in indoor environment for short intervals of time.

Section III

Mathematical Modeling and Numerical Study of the Properties of Woven Structures with Respect to Thermophysiological Comfort

12

Mathematical Modeling and Numerical Simulation of Air Permeability and Heat Transfer through Woven Macrostructures: State of the Art

12.1 Introduction

The simulation of transport processes in porous materials—in particular, in through-thickness direction of textile macrostructures—is a subject of research studies in recent decades, but it is particularly developed with the progress of the computer resources and rapid development of CFD (computational fluid dynamics). However, the experimental assessment of air permeability and heat transfer continues to be the most reliable and applied way for evaluation of the heat and mass transfer processes through textile macrostructures.

Currently, the possibilities for forecasting the air permeability and heat transfer through textiles include the following:

- *Empirical (regression) dependences*, obtained on the basis of statistical analyses of experimental results. The disadvantage of this method is that the regression equations rarely exceed the scope of the study, in which they are obtained, and do not allow significant extrapolation. This approach was used in Chapters 6 through 10.

- *Analytical equations* that, unlike the CFD, do not require the use of specific software and computing power. The Hagen–Poiseuille law for laminar flow in pipes and channels, which can be used to determine the flow through the pores of a textile structure, is of this type. The advantage of the method is the relative simplicity and ability of the computation to be done with software packages for numerical calculations (MS Excel, Julia, GNU Octave, etc.). The disadvantages of the method are in the limitations of its application (i.e., the Hagen–Poiseuille law is only valid for laminar flows). This approach is used in Chapter 13.

- *Advanced modeling tools* like software packages with general application (commercial software packages) or specialized software codes for solving particular computational problems. The methods of CFD are often used for simulation of heat and mass transfer processes. Modeling of air permeability and heat transfer in through-thickness direction of woven macrostructures is presented in Chapters 14 and 15.

12.2 Air Permeability Modeling

The assessment of the impact of textiles on the microenvironment around the human body, as already discussed in Section I, is reduced also to the evaluation of the ability of the textile macrostructure to retain an air layer around the skin or between the textile layers in the clothing system and to maintain the temperature of the air layer so that the surface temperature of the body to remain constant irrespective of the change of the temperature of the environment. Therefore, in the last four decades, a number of authors and groups researched the problem of the impact of the textile structure on its ability to retain or provide air transport: Dhingra and Postle (1977), Kullman et al. (1981), Paek (1995), and others. The effect of porosity on the fluid transfer was studied in the works of Hsieh (1995) and Delerue et al. (2003). Modeling of the textile structure and its air permeability was presented in the research by Gooijer et al. (2003a, 2003b), Belov et al. (2004), and Verleye et al. (2007). Tokarska (2004) simulated air permeability of fabrics, applying neural networks. Later on, Nazarboland et al. (2007) used CFD and FLUENT software package to study the filtering ability of woven structures. Wang et al. (2007) focused on the simulation of fluid flow in a single pore of a woven macrostructure, made of monofilaments. Grouve et al. (2008) presented a specialized CFD code for calculating the permeability of composite structures in transversal direction. Similarly, Nabovati et al. (2010) used Lattice Boltzmann method for studying the transport of fluids in the transverse direction of textile structures, woven from polyfilaments. Mao (2009) presented a model of the air permeability of two types of nonwovens, the structure of which included areas with low and high permeability, as is the case of some of the woven macrostructures.

Depending on the way of presentation of the woven macrostructure, the air permeability simulation of woven textiles is carried out in the following three main directions:

- Simulation of flow through a set of holes
- Simulation of flow through porous media of randomly distributed fibers
- Simulation of flow around a set of objects (bodies)

These ways of representation of the interaction between the flow and the woven macrostructure are applied in both analytical models and computer modeling methods.

Modeling of flow through a set of holes is the oldest and most commonly used method. The diversity in the studies is related primarily to the representation of the macrostructure. Great part of the research studies are devoted to attempts to simplify the geometry and the simulation itself through an idealized representation of the structure of woven textiles, made of mono- and polyfilaments, as a net. Another part of the publications considered a simplified, structured geometry, usually limited to one repeat of the fabric. In more recent publications, finite element method was applied in order to complicate this geometry and to get closer to the real way of interaction between warp and weft threads in the woven macrostructure. However, the idea of simulation of a *unit cell* or a single repeat of the weave (in general, plain weave or twill) only was kept.

Analytical model of this type was presented in Kulichenko and Langenhove (1992). They reported equations for air transport through the pores of both macro- and mesostructures. However, the model evaluated good enough only the air permeability of the macrostructure.

Cai and Berdichevsky (1993) studied the air permeability of a fiber bundle, dealing with the textile microstructure. The fibers' arrangement was considered to be the major problem in simulating the yarn structure and hence the simulation of the macrostructure. The authors summarized the existing models for fluids transfer in both through-thickness and in-plane directions and found that there were only two models for through-thickness permeability of the textiles: that of Gutowski et al. (1987) and Lain and Kardos (1988). The authors used ANSYS and FIDAP software packages for 2D simulation of flow through a bundle of fibers arranged in five different ways, based on a flow around body treatment. The results showed that the bulk density of the yarn was not a sufficient indicator to predict the air permeability, as the fiber arrangement in the yarn influenced the air flow movement.

Bruschke and Advani (1993) also studied the flow through a porous structure, presenting the interaction as a flow around an ordered set of fibers. They developed a model for porous media with 40% porosity, which, according to the authors, provided a good match with numerical results. The authors assumed that above the value of 40% the fibers stayed far enough from each other so as a fiber to disturb the flow around an adjacent fiber. However, that result contrasted with the findings of Cai and Berdichevsky (1993).

Phelan and Wise (1996) developed a model of a flow in the transverse direction of a rectangular ordered bundle of fibers with elliptical cross section. They studied the shape of the fibers and the effect of the permeability of the space between the fibers on the total permeability of the fiber medium. Later, Kulichenko (2005) presented a model of gases in porous media, assuming that the flow is laminar. The model was verified with experimental results for the air permeability of nonwovens, that is, medium with randomly distributed fibers.

Gooijer et al. (2003a and 2003b) presented analytical models for predicting the flow resistance of fabrics, made of monofilament and polyfilaments. They improved the analytical model of Rushton and Griffiths (1971) as it concerned the shape of the pores, adapting the model of Backer (1951). It is debatable, however, the importance of the geometry of the individual pore, presented by the model of Gooijer et al. (2003a), as in their next paper Gooijer et al. (2003b) concluded that the polyfilaments in the woven macrostructure were subject of deformation and could not be modeled as cylinders, which considerably reduced the importance of the complicated model of the geometry of the pores. A second important conclusion, which the authors made, was that in *open structure* woven textiles, that is, with lower warp and weft density, the substantial part of the fluid flow passed through the pores, and the percentage of flow through the mesostructure (the pores within the yarns) was negligible.

Saldaeva (2010) used a combination between the analytical model of Kulichenko and Van Langenhove (1992) for a description of the fluid flow through the pores of the woven macrostructure and the model of Gebart (1992) for a description of the fluid flow through the mesostructure. Along with this, the author performed a computer simulation of the through-thickness air permeability of woven macrostructures. Such an approach was applied by Xiao (2012), who developed an analytical model based on the Darcy law for laminar flows with very low Reynolds numbers. Experimental data were used for verification of the theoretical calculations.

Verleye et al. (2009) has indicated one of the essential shortcomings of the analytical models: although giving good similarity with experimental results, they are limited in nature and can hardly be used in the variety of applications of fluid flows, passing through textile macrostructures. The same was the conclusion of Kulichenko (2005), who pointed that the simplifications and material constants limited the application of analytical models for air permeability modeling.

Therefore, the computer simulations of fluid flows, based on fundamental laws of fluid mechanics, have far greater scope and perspective.

The advantage of CFD as a whole is that it allows the simulation of any process that involves fluids. CFD is used as a tool for predicting the air permeability of textiles, but, like in the case of experimental studies—due to the wide variety of structural, geometric, and mass characteristics of woven fabrics with different applications—the published studies are related to specific tasks and applications.

For simulation of the air permeability of textile structures, the following two approaches are generally used:

- Based on the system of Navier–Stokes partial differential equations
- Based on the lattice Boltzmann method

The advantages and disadvantages of the two approaches are mainly related with the grid, the *ease* in the preprocessing of complex geometries and the required central processing unit (CPU) time (Verleye et al. 2009). In general, unlike many CFD methods, lattice Boltzmann method does not solve the Navier–Stokes equations, but models the flow of a Newtonian fluid, consisting of fictive fluid particles that move and collide in a discrete lattice mesh. The movement of fluid particles is described by the kinetic Boltzmann equation, while the Bhatnagar–Gross–Krook operator describes the particles collision (Guo and Shu 2013).

CFD software packages like FLUENT, CFX, and PHOENICS are used for simulation of fluid flows and particularly can be applied for modeling of air permeability of textiles. Specialized software codes like NaSt3DGP and FlowTex are also mentioned in the literature, together with packages for simulation of both textile and composite structures (TexGen and WiseTex).

Sobera et al. (2004) investigated the heat and mass transfer in fabrics for protective clothing. Special simulation approach was used: the whole macrostructure was treated as porous continuum, through which the fluid flowed. The simulation was based on the Reynolds-averaged Navier–Stokes equations, solved with the use of FLUENT software package. The authors found that their approach gave sufficiently accurate results, particularly in terms of heat transfer. Direct numerical simulation method was also used to prove that the flow through the mesostructure was laminar and Darcy's law was valid for it.

Verleye (2008) applied the finite differences method for discretization of the Navier–Stokes partial differential equations (Navier–Stokes PDEs), using the NaSt3DGP open software code. However, the primary method used was the Lattice Boltzmann method and FlowTex software package. The permeability of composite and nonwoven materials in lateral direction was simulated as FlowTex is not suitable for prediction of through-thickness permeability.

Saldaeva (2010) performed a computer simulation of the air permeability of textile macrostructures in through-thickness direction. Only one *cell* of the woven macrostructure was modeled with the TexGen software package. The air permeability simulation was performed with CFX CFD software package. The analysis of the numerical results showed that the author did not simulate the air flow through the mesostructure, but only through the voids of the woven macrostructure.

Similar is the work of Xiao (2012), as it concerns the preprocessing and the simulation of the air permeability. The author compared the numerical results, obtained by using CFX software package, with the results of calculations, based on the analytical model of Gebart (1992) and found that the difference between the two applied methods was up to 20%.

The analysis of the literature shows that despite the significant number of research publications, the reports on simulation of air permeability using CFD for solving the Navier–Stokes equations are few. The simulation method presented in the next chapter has not been applied so far in the literature. In essence, the method refers to modeling of a flow through a set of openings, but their number is not an infinite one as it is in the case of nets or perforated plates. The number of openings is not approximated either to the woven repeat or to one cell unit, describing the interaction between two warp and weft threads.

12.3 Heat Transfer Modeling

Both analytical models and computer simulation are applied for prediction of the heat transfer through a textile layer. The problem includes again two tasks, which determine the diversity in research: building of the geometry of the textile macrostructure (particularly a woven one) and a selection or development of a mathematical model.

The modeling of heat transfer through porous fibrous barrier has attracted the attention of many researchers in recent decades. Objects of different studies are paper products, building materials, and, much later, textiles and clothing. The research in the field was probably initiated by Henry (1939), who described the diffusion of moisture in cotton bales. Later on, he studied the transfer of heat and moisture through a garment (Henry 1948). David and Nordon (1939) improved the Henry's model: they studied the transfer of heat and moisture in wool and avoided certain simplifications introduced by Henry (1939). Since then various models have been developed, some of which with general validity, others—for specific applications.

One of the well-known analytical models that described the heat transfer through textile fibers is the model of Farnworth (1983). The author developed a combined model for the heat transfer by conduction and radiation in fiber webs of various synthetic materials, and mixtures of down and feathers. He found that in these substances convection was not used as a way for heat transfer, with the exception of the case of a fiber web with very low bulk density. Classical thin textiles, however, were not included in the study.

The model of Li and Holcombe (1992) evaluated the joint transfer of heat and moisture in woolen fabrics. It was a two-step model that used different equations for description of the sorption processes. The authors applied the method of finite differences to obtain the numerical results. The model had good agreement with experimental results and other analytical models. It was not clear whether the model was applicable to macrostructures from other types of fibers.

The model of Le and Ly (1995) for heat transfer in case of forced convection was also discussed in the literature. However, forced convection could not be connected with the indoor environment. At the same time, Ghali et al. (1995) proposed a combined model for the transfer of heat and moisture in knitted macrostructures. It was difficult to assess the applicability of this model for woven macrostructures as it was based on a geometric model of the loop structure.

Besides the analytical models of heat transfer through a single textile layer, models have been developed for multilayer systems as well: Fohr et al. (2002), Ghali et al. (2002), and Wang et al. (2008). The model of Wang et al. (2008) described a multilayer system (clothing) as composed of two continuums and a dispersed phase, thus providing equations for heterogeneous materials. The authors determined that the model was suitable for prediction of the heat transfer through composite materials, but the results from the calculations were not verified with experimental data.

The computer simulation of heat transfer through thin textile layers is a modern, developing area, but the publications on this topic are still few, as commented by Saldaeva (2010) and Duru Cimilli et al. (2012).

Hossain et al. (2005) used FLUENT CFD software package for predicting the heat transfer through a porous layer of fibers (web), using Navier–Stokes PDEs and the energy equation. The authors investigated the effect of the air velocity, porosity, and thickness of the web on different technological characteristics—that is, the temperature of the web and the melting temperature of the fibers.

Instead of the control volumes method, used in the modeling of Hossain et al. (2005), Cui and Wang (2009) applied finite differences method as a method for discretization of PDEs, studying the heat transfer through fiber web at a variable air velocity. They found good agreement between theoretical and experimental results, obtained in a climatic chamber. Kothari and Bhattacharjee (2008) used CFD simulation for modeling of thermal resistance of woven textiles. They developed a mathematical model for predicting the heat transfer through conduction and convection. The second part of their study (Bhattacharjee and Kothari 2008) focused on the transfer of heat by convection and CFD simulations were performed. The authors concluded that the agreement between the results of the analytical model, computer simulation and experimental measurements was very good.

CFD was used in the research of Duru Cimilli et al. (2012), where the heat transfer by convection in knitted fabrics was studied. The virtual model, a thread with three loops, was built with CATIA V5R16 specialized software and meshed in GAMBIT. The computer simulation was performed with FLUENT, using a mathematical model that included the energy equation and a transport equation for the natural convection. The authors reported a very good match between experimental and numerical results for convective heat flux and concluded that the applied approach, including the control volumes method for discretization, could be successfully used for CFD simulation of heat transfer in knitwear.

The literature review has shown that the heat transfer through thin, single-layer textiles with woven macrostructure is poorly investigated by means of CFD. More attention was paid to nonwoven macrostructures, composite materials, and multilayer structures. The method for simulation of heat transfer through woven macrostructure presented in the next chapter has not been described so far in the literature.

In Section III of the book, results from mathematical modeling and numerical simulation of air and heat transfer through woven macrostructures are presented as follows:

- Mathematical modeling and numerical simulation of the air permeability of woven macrostructures by applying the Hagen–Poiseuille law (Chapter 13)
- Mathematical modeling and numerical simulation of the air permeability of woven structures by CFD (Chapter 14)
- Mathematical modeling and numerical simulation of heat transfer through woven structures by CFD (Chapter 15)

12.4 Summary

The state of the art in the field of mathematical modeling and numerical simulation of air permeability and heat transfer through woven textiles was presented.

13

Simulation of Air Permeability by Using Hagen–Poiseuille Law

Woven fabrics are produced by interlacing of warp threads (located along the fabric) and weft threads (square to the warp threads). The threads are separated from each other by technological reasons: the reed beating and the position of the warp on the weaving machine (Lord 1959). As a result, voids are formed between the warp and weft threads.

The voids in all type of textile fabrics (woven, knitted, and nonwoven) play an important role in a variety of consumer and industrial applications, related with thermal comfort and insulation efficiency, barrier performance, and so on (Epps and Leonas 1997; Angelova and Nikolova 2009). There are several applications of woven fabrics, where the permeability of the structure must be known and maintained into a specific range. Low permeability fabrics are used for production of air bags, for example. Air bags are adapted to be inflated instantaneously with high-pressure gas in the event of a collision, so as to prevent the movement of the passenger for safety reasons. For this purpose, the fabric has to be with low air permeability and, in addition, to be strong, flexible, thin, and lightweight.

Precision woven fabrics with controlled permeability are used for medical filters or separation media. Woven structures are also used in medical manipulations and apparatuses when these structures have to guarantee unidirectional paths. Fabrics with such an application type are made of monofilaments, polyfilaments, or combination of both. Different types of densities, porosity, and weaves are used.

The chapter presents a numerical procedure for prediction of the air permeability of woven structures, based on Hagen–Poiseuille law. The pores of the woven macrostructure are considered to be microtubes, whose dimensions are described by using specific characteristics of the woven structure and its components: linear density of the warp and weft threads, number of warp and weft threads per centimeter, construction phase of the yarn interlacing, shape of the voids, thickness of the fabric, and so on.

The presented procedure is expected to allow relatively fast assessment of the air permeability of the macrostructure. The procedure was verified with data from literature on the air permeability of 28 samples of worsted fabrics. Than it was used for prediction of the air permeability of the cotton fabrics, studied in Chapter 7, and the theoretical results were compared with the experimental values of the air permeability coefficient B_p of the woven macrostructures.

13.1 Applicability of the Hagen–Poiseuille Law

Hagen–Poiseuille law allows to determine the pressure drop along a cylindrical tube. It is worked out on the following conditions (Johnson 1998):

- The fluid is viscous and incompressible.
- The fluid flow is laminar.
- The dimensions of the tube opening remain the same alongside the tube and the length is greater than the tube diameter.

The flow rate Q_a, m^3/s, through a circular pipe is determined by the law:

$$Q_a = \frac{\Delta P_p \pi d_p^4}{128 \mu L_p} \tag{13.1}$$

where:
ΔP_p is the pressure difference between inlet and outlet of the pipe, Pa
d_p is the inner diameter of the pipe, m
μ is the dynamic viscosity, Pa.s
L_p is the length of the pipe, m

Hagen–Poiseuille law can be applied for prediction of the through-thickness air permeability of woven structures, as in the case of absence of forced convection and indoor environmental conditions, the air flow through the voids of the fabric is viscous, incompressible, and laminar (Angelova 2011). However, the following assumptions have to be taken into account:

- All interstices between the threads in the fabric have the same size.
- The void size does not change along its length *l*.
- The void length is constant and equal to the average thickness δ of the fabric.

13.2 Mathematical Model for Calculation of Air Permeability

The application of the Hagen–Poiseuille law requires the interrelation between the flow field through a void and the fabric parameters to be logically described through geometric characteristics of the meso- and macrostructures. Similar approach was used in the work of Xu and Wang (2005).

The systematic algorithm for calculation includes the following steps (Angelova 2011).

13.2.1 Calculation of the Number of Voids n_p in the Woven Macrostructure

The following expression is applied for the calculation of the number of voids in the woven macrostructure:

$$n_p = (n_{wa} - 1)(n_{wf} - 1) \tag{13.2}$$

where:
n_{wa} is the number of warp threads
n_{wf} is the number of the weft threads in a particular area

If the area of the calculation is 10×10 cm, instead of n_{wa} and n_{wf}, the warp P_{wa} and weft density P_{wa}, threads/dm, can be used.

13.2.2 Calculation of the Diameter of the Threads

Dependences between the linear density and diameter of the thread are used as follows:

$$d_{wa} = 0.0357 \sqrt{\frac{Tt_{wa}}{\gamma_m}} \tag{13.3}$$

$$d_{wf} = 0.0357 \sqrt{\frac{Tt_{wf}}{\gamma_m}} \tag{13.4}$$

where:
d_{wa} and d_{wf} are the diameters of warp and weft threads, respectively, m
Tt_{wa} and Tt_{wf} are the respective linear density, tex
γ_m is the mean density of the thread, g/mm³

13.2.3 Determination of the Pore Size

Figure 13.1 presents a 3D structure of a plain weave fabric. Each of the voids between the threads can be considered as a rectangular one with dimensions a and b, calculated from

$$a = \frac{L_s}{n_{wa}} - d_{wa} \tag{13.5}$$

$$b = \frac{L_s}{n_{wf}} - d_{wf} \tag{13.6}$$

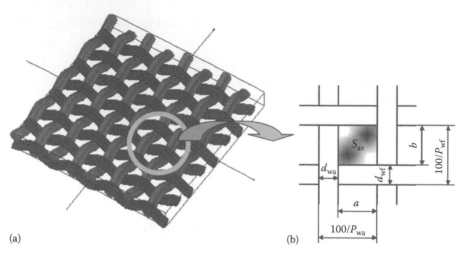

FIGURE 13.1
Structure of a void in a plain weave fabric (a) and geometrical dependencies (b).

where:

L_s is the length of the sample, m

When the warp density P_{wa} (threads/dm) and the weft density P_{wf} (threads/dm) are known, the dimensions of the voids can be also calculated as

$$a = \frac{100}{P_{wa}} - d_{wa} \tag{13.7}$$

$$b = \frac{100}{P_{wf}} - d_{wf} \tag{13.8}$$

13.2.4 Calculation of the Pore Area and Pore Perimeter

The average pore area S_{av}, m^2, is calculated as

$$S_{av} = a \cdot b \tag{13.9}$$

and the average perimeter P_{av}, m, as

$$P_{av} = 2(a + b) \tag{13.10}$$

13.2.5 Determination of the Porosity of the Macrostructure

The following expression can be used for the determination of the porosity of the macrostructure:

$$\varepsilon_f = \frac{n_p \cdot S_{av}}{S_f}, \% \tag{13.11}$$

where:

S_f is the area of the fabric sample, m^2

13.2.6 Calculation of the Thickness

If experimental data about the sample's thickness are available, this step can be avoided. Otherwise, empirical equations are used, like that of Kulichenko and Langenhove (1992), which is valid for a balanced woven structure (5th phase of construction) (Figure 13.2):

$$\delta = 1.5 d_o, m \tag{13.12}$$

13.2.7 Calculation of the Hydraulic Diameter of the Void

The hydraulic diameter d_h of the void is calculated as follows:

$$d_h = \frac{4 S_{av}}{P_{av}}, m \tag{13.13}$$

13.2.8 Determination of the Mean Velocity

If the head losses are taken into account, the pressure drop can be calculated from the expression

$$\Delta P_p = \lambda_p \frac{c_v^2}{2} \frac{\delta}{d_h} \rho_a \tag{13.14}$$

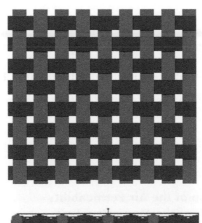

FIGURE 13.2
Balanced woven macrostructure.

where:

ρ_a is the air density for the respective temperature, kg/m^3

The head losses coefficient λ_p for a laminar flow in a pipe is

$$\lambda_p = \frac{64}{Re} \tag{13.15}$$

where Re, the Reynolds number, is defined as

$$Re = \frac{c_v d_h}{\nu} \tag{13.16}$$

where:

ν is the kinematic viscosity, m^2/s

Taking into account Equations 13.15 and 13.16, the mean velocity of the air through a pore with a hydraulic diameter d_h is calculated from

$$c_v = \frac{d_h^2}{32\mu\delta}\Delta P, \text{m/s} \tag{13.17}$$

where:

the dynamic viscosity μ is 18.27×10^{-6} Pa.s

13.2.9 Calculation of the Hydraulic Area of the Pore

It is based on the calculation of the hydraulic diameter d_h:

$$S_{eq} = \frac{\pi d_h^2}{4}, \text{m}^2 \tag{13.18}$$

13.2.10 Determination of the Air Flow Rate through the Sample

The following equation is used for the determination of the air flow rate through the sample:

$$Q_a = \frac{n_p}{\varepsilon_f} S_{eq} c_v, \text{m}^3/\text{s} \tag{13.19}$$

13.2.11 Determination of the Air Permeability Coefficient B_p of the Macrostructure

$$B_p = \frac{Q_a}{S_f}, \text{m}^3/\left(\text{m}^2.\text{s}\right) \tag{13.20}$$

13.3 Theoretical Results

The Hagen–Poiseuille law enables multi-parametrical analysis of the process of air transfer through the woven textile to be done as a function of the characteristics not only of the macrostructure but also of the meso-structure. The presented mathematical model was used to determine the theoretical air permeability of 11 woven macrostructures, arranged in two groups:

- *Group 1*: The effect of the change of the weft density P_{wf} and the linear density of the weft threads Tt_{wf} on the air permeability in through-thickness direction is simulated, when keeping constant both the warp density P_{wa} and the linear density of the warp threads Tt_{wa}.
- *Group 2*: The effect of the change of warp P_{wa} and weft P_{wf} density on the air permeability in through-thickness direction is simulated, when keeping constant the linear density of both warp Tt_{wa} and weft Tt_{wf} threads.

Data for the simulation of the 55 samples from group 1 are summarized in Table 13.1. The warp density was constant (200 threads/cm), while different types of weft filling density were used: 11 variations from 200 to 400 threads/cm. The linear density of the weft threads was changed as well: five variants from 50 to 16.7 tex. The warp threads linear density was constant ($Tt_o = 20$ tex). It was accepted that the woven structure was balanced (5th phase of construction).

Example results from the simulation of the air permeability coefficient B_p of the macrostructures from group 1 are shown in Figure 13.3. The analysis of the results showed that the increase of the weft density led to a decrease of B_p. Similar was the effect of the increase of the linear density of the weft threads.

The results were logical, as both the higher number of picks per centimeter and thicker weft yarns led to a production of tighter fabrics. The void dimensions decreased, resulting in a decrease of the air permeability coefficient. Therefore, the sample with the thickest weft yarn (50 tex) and highest weft density (40 threads/cm) showed the lowest air permeability.

For the simultaneous evaluation of the effect of both warp and weft density, a new set of 55 woven samples was simulated (group 2). In this case, the linear density of warp and weft threads was constant, 50 tex. Data for the simulated macrostructures are summarized in Table 13.2.

The results of the theoretical air permeability of the samples are shown in Figure 13.4. The graphs presented the effect of the simultaneous change of the threads density in both directions of the macrostructure on the air

TABLE 13.1

Simulated Woven Macrostructure, Group 1

Sample	P_{war} threads/dm	P_{wft} threads/dm	Tt_{war} tex	Tt_{wft} tex	Sample	P_{war} threads/dm	P_{wft} threads/dm	Tt_{war} tex	Tt_{wft} tex	Sample	P_{war} threads/dm	P_{wft} threads/dm	Tt_{war} tex	Tt_{wft} tex
1	200	200	20	16.7	20	200	360	20	20	39	200	300	20	33.3
2	200	220	20	16.7	21	200	380	20	20	40	200	320	20	33.3
3	200	240	20	16.7	22	200	400	20	20	41	200	340	20	33.3
4	200	260	20	16.7	23	200	200	20	25	42	200	360	20	33.3
5	200	280	20	16.7	24	200	220	20	25	43	200	380	20	33.3
6	200	300	20	16.7	25	200	240	20	25	44	200	400	20	50
7	200	320	20	16.7	26	200	260	20	25	45	200	200	20	50
8	200	340	20	16.7	27	200	280	20	25	46	200	220	20	50
9	200	360	20	16.7	28	200	300	20	25	47	200	240	20	50
10	200	380	20	16.7	29	200	320	20	25	48	200	260	20	50
11	200	400	20	16.7	30	200	340	20	25	49	200	280	20	50
12	200	200	20	20	31	200	360	20	25	50	200	300	20	50
13	200	220	20	20	32	200	380	20	25	51	200	320	20	50
14	200	240	20	20	33	200	400	20	25	52	200	340	20	50
15	200	260	20	20	34	200	200	20	33.3	53	200	360	20	50
16	200	280	20	20	35	200	220	20	33.3	54	200	380	20	50
17	200	300	20	20	36	200	240	20	33.3	55	200	400	20	50
18	200	320	20	20	37	200	260	20	33.3					
19	200	340	20	20	38	200	280	20	33.3					

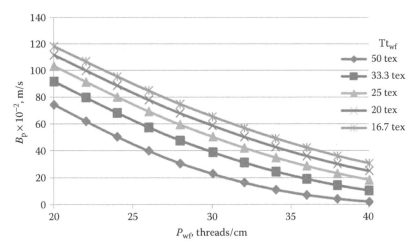

FIGURE 13.3
Theoretical air permeability of different type of woven macrostructures with changeable weft density and linear density of the weft threads.

permeability coefficient B_p. The strongest change in B_p appeared for the highest weft density and the lowest warp density: $P_{wa} = 20$ threads/cm. The increment of the warp density to $P_{wa} = 25$ threads/cm also caused a change of B_p, but in a much smaller range: from 0.53 to 1.0 m/s.

The simulation on Figure 13.4 clearly showed that the effect of the change of the weft density on the air permeability was very small for warp density $P_{wa} = 35$ threads/cm and $P_{wa} = 40$ threads/cm. Moreover, the values for B_p of the simulated macrostructures were very similar for $P_{wf} > 36$ threads/cm, despite the change of the warp density P_{wa}.

From a practical perspective, this means that the optimal combination of warp and weft density had to be found to provide a particular coefficient of air permeability. This is especially important for the production of woven macrostructures with high density and low air permeability: the implicit increase of the threads density in both directions may result in a greater consumption of material than is necessary to achieve the desired (low) permeability.

Figure 13.5 presents the theoretical results of the air permeability as a function of the porosity ε_f. It was reasonable to expect that the high porosity would provoke high levels of air permeability. Conversely, the macrostructure with zero porosity was expected to be impermeable: a requirement, typical for woven fabrics used as sails, for example.

The analysis of the results showed that warp density strongly influenced the porosity—tighter fabrics, produced with higher number of threads/cm, had lower values of air permeability. The curves showed that the warp density could hardly influence the air permeability of the macrostructures with

TABLE 13.2

Simulated Woven Macrostructure, Group 2

Sample	P_{wa} threads/dm	P_{wr} threads/dm	Tt_{wa} tex	Tt_{wr} tex	Sample	P_{wa} threads/dm	P_{wr} threads/dm	Tt_{wa} tex	Tt_{wr} tex	Sample	P_{wa} threads/dm	P_{wr} threads/dm	Tt_{wa} tex	Tt_{wr} tex
1	200	200	50	50	20	250	360	50	50	39	350	300	50	50
2	200	220	50	50	21	250	380	50	50	40	350	320	50	50
3	200	240	50	50	22	250	400	50	50	41	350	340	50	50
4	200	260	50	50	23	300	200	50	50	42	350	360	50	50
5	200	280	50	50	24	300	220	50	50	43	350	380	50	50
6	200	300	50	50	25	300	240	50	50	44	350	400	50	50
7	200	320	50	50	26	300	260	50	50	45	400	200	50	50
8	200	340	50	50	27	300	280	50	50	46	400	220	50	50
9	200	360	50	50	28	300	300	50	50	47	400	240	50	50
10	200	380	50	50	29	300	320	50	50	48	400	260	50	50
11	200	400	50	50	30	300	340	50	50	49	400	280	50	50
12	250	200	50	50	31	300	360	50	50	50	400	300	50	50
13	250	220	50	50	32	300	380	50	50	51	400	320	50	50
14	250	240	50	50	33	300	400	50	50	52	400	340	50	50
15	250	260	50	50	34	350	200	50	50	53	400	360	50	50
16	250	280	50	50	35	350	220	50	50	54	400	380	50	50
17	250	300	50	50	36	350	240	50	50	55	400	400	50	50
18	250	320	50	50	37	350	260	50	50					
19	250	340	50	50	38	350	280	50	50					

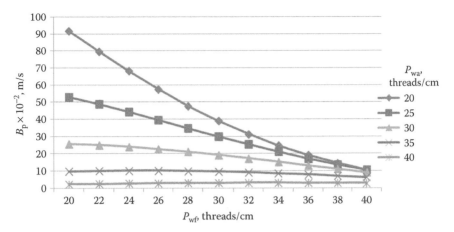

FIGURE 13.4
Theoretical air permeability of different type of woven structures with changeable weft density and linear density of the weft threads.

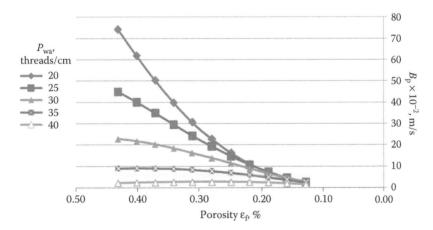

FIGURE 13.5
Theoretical air permeability of different type of woven macrostructures with changeable warp density and porosity.

low porosity ($\varepsilon_f < 20\%$). Tightly woven fabrics (35 and 40 threads/cm) showed low permeability that was almost constant for all values of porosity. At the same time, the warp density had a significant influence on B_p of the macro-structures with porosity of $\varepsilon_f > 20\%–25\%$.

In conclusion, the mathematical model, based on the Hagen–Poiseuille law, allowed to forecast the air permeability of woven macrostructures. The results from the simulation of 110 woven samples were logical and gave the possibility to make qualitative assessments of the impact of the characteristics of the meso- and macrostructures on the air permeability in through-thickness direction of the samples.

13.4 Verification of the Model

Xu and Wang (2005) showed details from the simulation of the air permeability of 28 worsted fabrics, applying the Hagen–Poiseuille law. The results from the study were used for the verification of the described mathematical model: the theoretical results for the hydraulic pore diameter d_h, the fabric cover factor E_s, and the air permeability coefficient B_p of the samples were compared with the results from the work of Xu and Wang (2005).

The results from the calculation of the hydraulic diameter d_h of a single pore of the woven macrostructures are summarized in Table 13.3. The comparison with the results of Xu and Wang (2005) showed that there was big coincidence with the data from literature. However, 8 of the cases (samples 4, 6, 8, 10, 22–24, and 26) demonstrated significant differences in the values for the pore diameters. These discrepancies were difficult to be explained on condition that the geometric model for calculating the pore area was identical. The highest relative error of 206% was found for samples 10 and 23.

By analogy the intermediate results for the calculation of the cover factor were compared in Table 13.4. The cover factor values were almost identical with the data of Xu and Wang (2005) and the only exceptions were the E_s values for samples 21 and 26: the relative error for sample 21 was 19% and for sample 26 was 4%.

TABLE 13.3

Comparison between the Results for the Hydraulic Diameter of the Pores from the Model and the Study

Hydraulic Diameter of the Pores d_h, mm													
Sample	1	2	3	4	5	6	7	8	9	10	11	12	13	14
d_h (Xu and Wang, 2005)	0.078	0.033	0.184	0.033	0.049	0.082	0.089	0.035	0.062	0.085	0.057	0.094	0.028	0.115
$d_{h,\,model}$	0.082	0.033	0.184	0.084	0.050	0.106	0.089	0.014	0.061	0.260	0.056	0.094	0.029	0.115
Sample	15	16	17	18	19	20	21	22	23	24	25	26	27	28
d_h (Xu and Wang, 2005)	0.082	0.086	0.010	0.051	0.101	0.070	0.103	0.107	0.070	0.032	0.041	0.023	0.084	0.074
$d_{h,\,model}$	0.082	0.086	0.010	0.051	0.101	0.073	0.112	0.065	0.214	0.061	0.040	0.102	0.082	0.080

Source: Xu, G. and Wang, F., *J. Ind. Text.*, 34, 243–254, 2005.

TABLE 13.4

Comparison between the Results for the Fabric Cover Factor from the Model and the Study

Fabric Cover Factor, %													
Sample	1	2	3	4	5	6	7	8	9	10	11	12	13	14
E_s (Xu and Wang, 2005)	78.70	88.40	74.00	89.40	90.50	91.90	85.80	92.30	91.00	81.80	93.50	80.10	88.30	89.10
$E_{s,\,model}$	79.16	88.99	74.65	84.85	90.99	92.47	86.40	92.82	91.56	82.46	94.11	80.36	88.33	89.18
Sample	15	16	17	18	19	20	21	22	23	24	25	26	27	28
E_s (Xu and Wang, 2005)	80.70	87.00	91.70	90.80	84.30	76.20	79.90	79.90	78.90	98.20	104.0	112.6	89.30	90.70
$E_{s,\,model}$	80.74	87.50	91.70	91.88	85.00	79.25	95.17	80.55	79.51	101.85	103.97	108.3	90.50	91.26

Source: Xu, G. and Wang, F., *J. Ind. Text.*, 34, 243–254, 2005.

TABLE 13.5

Comparison between the Results for the Air Permeability Coefficient from the Model and the Study

Air Permeability Coefficient $B_p \times 10^{-3}$, m/s													
Sample	1	2	3	4	5	6	7	8	9	10	11	12	13	14
B_p (Xu and Wang, 2005)	90.7	1.5	629.6	1.5	23.0	77.5	36.8	1.8	20.5	34.1	42.1	194	3.54	313.7
$B_{p,\,model}$	114.4	1.5	627.6	50.2	23.5	217.3	36.8	0.05	19.5	2944.1	40.9	194	4.00	367
Sample	15	16	17	18	19	20	21	22	23	24	25	26	27	28
B_p (Xu and Wang, 2005)	118.7	178.9	0.01	31.1	283.7	12.13	569.1	425	43.50	7.74	10.70	0.55	39.4	70.9
$B_{p,\,model}$	118.3	179.7	0.01	31.0	282.9	13.95	794.4	77.0	3717	99.71	10.03	201	35.07	93.05

Source: Xu, G. and Wang, F., *J. Ind. Text.*, 34, 243–254, 2005.

Finally, the results for the air permeability coefficient B_p were compared in Table 13.5 and Figure 13.6. The analysis of the results led to the conclusion that in the majority of the samples, the results of the described model matched or were similar to those of the publication of Xu and Wang (2005). Deviations were due to the accumulation of error, mainly resulting from the differences in the values, obtained for the hydraulic diameter of the pores. This could explain the resulting peaks for examples 10 and 23 on Figure 13.6. The coincidences were however enough, with a small relative error, to conclude that the proposed mathematical model, based on the Hagen–Poiseuille law, and the mathematical model applied by Xu and Wang (2005), based on the same law, were very close. The verification allowed the model presented to be applied for numerical simulation of the air permeability of other woven macrostructures.

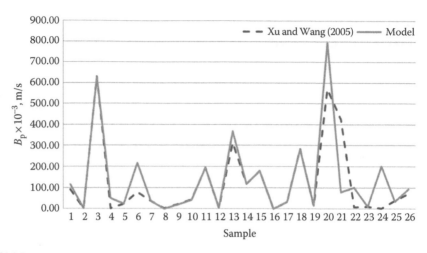

FIGURE 13.6
Comparison between the results for the air permeability coefficient from the model and the study. (Data from Xu, G. and Wang, F., *J. Ind. Text.*, 34, 243–254, 2005.)

13.5 Numerical Simulation of the Air Permeability of Textiles for Clothing and Bedding

13.5.1 Numerical Procedure and Results

The 14 woven macrostructures for clothing and bedding, whose air permeability was experimentally investigated in Chapter 7, were used to simulate the air permeability by using the described mathematical procedure, based on the Hagen–Poiseuille law. The simulation of the air permeability was carried out using the data of characteristics of the meso- and macrostructures of the finished fabrics, listed in Tables 7.3 and 7.4 (Chapter 7).

The results of the simulated mean velocity c_v of the air flow through a single pore of the particular sample, with a pressure difference of 100 Pa between the two sides of the sample, are shown in Figure 13.7. The highest velocity was obtained for samples 1 and 2, due to the high porosity of their macrostructure in comparison with other samples, and the resulting values of the hydraulic diameter of the single pore.

The velocity of the flow determines the Reynolds number and a verification of the conditions for laminar flow was performed. The results are shown in Figure 13.8. Obviously, the Reynolds number for samples 1 and 2 significantly exceeded the limit value of Re = 2320 for a laminar flow in a pipe, and one of the conditions for the application of the law of Hagen–Poiseuille was not fulfilled for these samples. For the remaining samples 3–14, however, the Reynolds number Re << 2320 and the air flow through a pore of the simulated macrostructures was assessed as laminar.

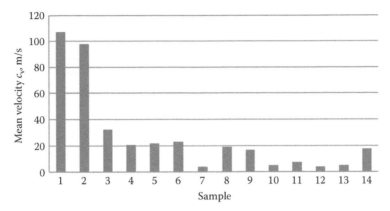

FIGURE 13.7
Results from the simulation: mean velocity through one pore.

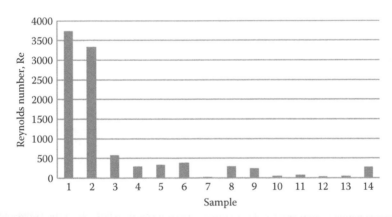

FIGURE 13.8
Results from the simulation: Reynolds number.

Figure 13.9 visualizes the results for the flow rate through a single pore, which logically follow the curve from Figure 13.7.

13.5.2 Verification of the Results

The results for the air permeability coefficient B_p of the simulated macrostructures are presented in Figure 13.10. The same graph shows the experimental results for B_p (from Table 7.7). The comparison allowed the presented mathematical procedure to be verified with a new set of experimental data.

The graph clearly showed the deviation of the numerical from the experimental results. The calculated relative error ranged from 5% (absolute value) for

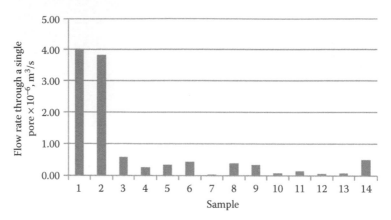

FIGURE 13.9
Results from the simulation: air flow rate through one pore.

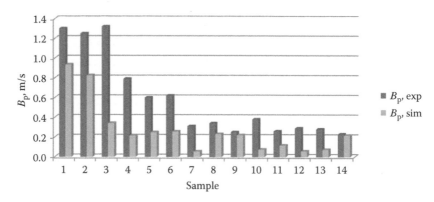

FIGURE 13.10
Comparison between numerical and experimental results for the air permeability coefficient B_p.

sample 14 and 13% for sample 9, to 412% for sample 10 and 463% for sample 7. It should be noted that the relative error for samples 1 and 2, where the flow through the pore was assessed as a turbulent one (see Figure 13.8) was 39% for sample 1 and 51% for sample 2.

The last, however, does not change the fact that an essential condition for the application of Hagen–Poiseuille law is not observed.

An important result from the simulation was that the numerical values for B_p were smaller than the experimental for all samples. This effect could be explained with the very idea of the application of the Hagen–Poiseuille law: it was used for the simulation of the air flow through the pores of the macrostructure only, but actually part of the flow passed through the pores of the mesostructure as well. That flow through the mesostructure was not taken into account.

The analysis of the literature data that concern the applicability of the Hagen–Poiseuille law for prediction of the air permeability of textiles showed that the discrepancies between the calculated and measured values by Xu and Wang (2005), for example, were of the same order as in Figure 13.10. Ogulata (2006) did not even try to compare the predicted results with the real air permeability of the investigated samples.

It is therefore logical to conclude that the use of the law of Hagen–Poiseuille for simulating the air permeability of woven structures in through-thickness direction does not give reliable results in quantitative sense. The causes must be sought mainly in the series of simplifications in the mathematical model as well as the physical nature and complexity of the task to simulate the air permeability of porous (particularly woven) macrostructure.

13.6 Summary

A mathematical model for calculation of the air permeability coefficient of woven macrostructures, based on the Hagen–Poiseuille law, was presented. The model was verified through simulation of the air permeability of 28 worsted fabrics. The comparison with literature data showed a good match, proving the feasibility of the proposed mathematical procedure.

Through a simulation of the air permeability of 110 woven macrostructures, it was found that the mathematical model, based on the law of Hagen–Poiseuille, could be applied for qualitative assessment of the impact of characteristics of the meso- and macrostructure of woven samples on their air permeability. The model would have practical importance during the phase of design of new macrostructures and textile items.

A simulation of the air permeability of 14 woven macrostructures for clothing and bedding was performed. The results from the prediction were compared with experimental results from Chapter 7. It was found that the quantitative coincidence between the predicted and measured values of the air permeability coefficient is quite low. Therefore, it was concluded that the model for calculation of the air permeability of woven macrostructures, based on the Hagen–Poiseuille law, is not suitable for reliable predictions.

14

Mathematical Modeling and Numerical Simulation of the Air Permeability of Woven Structures by CFD

Computational fluid dynamics (CFD) is a powerful modern tool for solving engineering problems of different essence, whose physical nature involves fluids. The investments in terms of developing specialized CFD codes or purchasing licensed CFD software packages and training to work with them are great, but CFD has significant advantages over the experimental studies, which can be summarized as follows:

- The numerical analysis is based on a correct mathematical model of the flow, represented by a system of nonlinear partial differential equations.
- The local parameters of the flow are predicted; thus, an extensive database that allows detailed analysis of the processes is obtained.
- The numerical results can be visualized, which greatly facilitates the analysis of the simulated process.
- The total costs for the numerical study of a specific task remain significantly lower than the costs of experiments.

Furthermore, CFD simulations allow to obtain data on the design phase of an item or a process, without a real prototype to be made or to study fluid processes, for which the experimental research is labor intensive or impossible.

Nowadays CFD codes or software packages involve three main elements: preprocessor, solvers, and post-processor.

FLUENT CFD software package was used in this part of the study with its preprocessor GAMBIT.

14.1 Methods for Modeling of Fluid Flows

The following methods can be applied for modeling of fluid flows (Wendt 1992; Wersteeg and Malalsekera 1995; Stankov 1998):

- *Engineering method*: It is based mostly on establishing empirical relationships. Unlike its great use in the research in the field of textiles, it is becoming less used tool for modeling of fluid flows and especially of turbulent flows, due to the availability of more effective research methods.

- *Method based on direct numerical simulation (DNS)*: It is related with the direct solving of the Navier–Stokes partial differential equations, which requires the density of the computational grid to be in the order of the Kolmogorov's length. This implies a very fine computational grid to be used, resulting in a system of a huge number of algebraic equations, which calls for the use of extremely powerful computers even for flows at low Reynolds numbers.

- *Method based on the Reynolds-averaged Navier–Stokes equations (RANS)*: It is the most used method to solve a wide range of complex engineering tasks, which is primarily preconditioned by the following:

  - The turbulent flows are sufficiently well described by the system of Reynolds equations.
  - The rapid development of computing promotes the intensive development of methods for solving partial differential equations.
  - The development of experimental research creates the necessary database for verifying the numerical results and optimizing the numerical experiments.

  The main problem in solving the system of Reynolds-averaged Navier–Stokes equations is that the system is open. Closing of the system of partial differential equations (PDEs) is done through different *turbulent models*.

- *Method based on the large eddy simulation (LES)*: It is built on the hypothesis of presentation of the turbulent flow as interaction of macrovortices, which contain in their volume a large number of microvortices. Generating the appropriate computing grid leads to an initial *neglect* of the microvortices and modeling only of the macrovortices with DNS. Microvortices are modeled through additional models based on RANS. The computing power necessary for this method is smaller than the necessary DNS, but applying LES for simulation of complex 3D flows requires powerful computing.

14.2 Mathematical Model

The system of partial differential equation used is the Reynolds-averaged Navier–Stokes equations that for an incompressible flow have the following form:

$$
\frac{\partial u}{\partial t} + u\frac{\partial u}{\partial x} + v\frac{\partial u}{\partial y} + w\frac{\partial u}{\partial z} = X - \frac{1}{\rho}\frac{\partial p}{\partial x} + \nu\left(\frac{\partial^2 u}{\partial x^2} + \frac{\partial^2 u}{\partial y^2} + \frac{\partial^2 u}{\partial z^2}\right)
$$
$$
-\rho\left(\frac{\partial \overline{u'}^2}{\partial x} + \frac{\partial \overline{u'v'}}{\partial y} + \frac{\partial \overline{u'w'}}{\partial z}\right)
$$

(14.1)

$$
\frac{\partial v}{\partial t} + u\frac{\partial v}{\partial x} + v\frac{\partial v}{\partial y} + w\frac{\partial v}{\partial z} = Y - \frac{1}{\rho}\frac{\partial p}{\partial y} + \nu\left(\frac{\partial^2 v}{\partial x^2} + \frac{\partial^2 v}{\partial y^2} + \frac{\partial^2 v}{\partial z^2}\right)
$$
$$
-\rho\left(\frac{\partial \overline{u'v'}}{\partial x} + \frac{\partial \overline{v'}^2}{\partial y} + \frac{\partial \overline{v'w'}}{\partial z}\right)
$$

(14.2)

$$
\frac{\partial w}{\partial t} + u\frac{\partial w}{\partial x} + v\frac{\partial w}{\partial y} + w\frac{\partial w}{\partial z} = Z - \frac{1}{\rho}\frac{\partial p}{\partial z} + \nu\left(\frac{\partial^2 w}{\partial x^2} + \frac{\partial^2 w}{\partial y^2} + \frac{\partial^2 w}{\partial z^2}\right)
$$
$$
-\rho\left(\frac{\partial \overline{u'w'}}{\partial x} + \frac{\partial \overline{v'w'}}{\partial y} + \frac{\partial \overline{w'}^2}{\partial z}\right)
$$

(14.3)

where:
  u, v, and w are the velocity components in the direction of x, y, and z ordinates, respectively
  u', v', and w' are the fluctuation velocities in the direction of x, y, and z ordinates
  X, Y, and Z are the components of the intensity of the body forces field, acting on the fluid in the direction of x, y, and z ordinates, respectively
  ρ is the fluid density, kg/m³
  P is the pressure, Pa
  ν is the kinematic viscosity, m²/s

The continuity equation is added to Equations 14.1 through 14.3:

$$
\frac{\partial u}{\partial x} + \frac{\partial v}{\partial y} + \frac{\partial w}{\partial z} = 0
$$

(14.4)

The tensor of the Reynolds stresses $P_{i,j}$ ($i, j = 1, 2, 3$) describes the additionally appeared in the PDEs turbulent stresses as follows:

$$P_{i,j} = \begin{vmatrix} -\rho\overline{u'}^2 & -\rho\overline{u'v'} & -\rho\overline{u'w'} \\ -\rho\overline{u'v'} & -\rho\overline{v'}^2 & -\rho\overline{v'w'} \\ -\rho\overline{u'w'} & -\rho\overline{v'w'} & -\rho\overline{w'}^2 \end{vmatrix} \tag{14.5}$$

The system of Reynolds PDEs is open: it contains 13 unknowns (10 if the Reynolds stresses tensor is symmetrical) and 4 equations. Two basic approaches are possible to close the system:

- Nine (six) additional transport equations have to be written for every turbulent stresses.
- The number of the unknowns is reduced by describing the Reynolds stresses through the averaged parameters of the flow, via the application of different turbulent models.

14.3 Turbulence Modeling

The nature and development of turbulent models determines the degree of applicability of the system of Reynolds partial differential equations for modeling of complex turbulent flows in the engineering practice. Turbulent models must be sufficiently complex to describe correctly the turbulence, but the complexity hampers the numerical realization.

The turbulent models can be divided into the following main groups (Stankov 1998):

- Reynolds stress models (RSM)
- Eddy viscosity models (EVM)
- Algebraic stress models (ASM)

In RSM, called *second-order closers*, the system of Reynolds PDEs is closed by adding transport equations for each of the turbulent stresses. ASM are a variety of RSM, because instead of additional transport partial differential equations for the Reynolds stresses (less significant for the flow) algebraic equations can be solved. So ASM appear to be a compromise between the drawbacks of EVM and complexity of RSM. In this sense, the RSM may be considered as algebraic RSM (actually ASM) and differential RSM (Stankov 1998).

EVM are based on the concept of turbulent eddy viscosity μ_t, expressing the interaction between the fluid eddies in the turbulent flow. The basic approach used is to reduce the number of unknowns in the Reynolds PDEs and bring them to a total of

- *4 unknowns*: For the models with zero additional equations.
- *5 unknowns*: For the models with one additional equation.
- *6 unknowns*: For the models with two additional equations.

The EVM with zero additional equations involve the models with *mixing length*. The models with one additional equation take into account the influence of the kinetic energy k of the turbulent pulsations on the turbulent eddy viscosity μ_t, and use and additional transport equation for the kinetic energy. The $k - \varepsilon$ model has acquired greatest popularity among the models with two additional equations: besides the transport equation for k, the model adds another one for the velocity of dissipation of the kinetic energy ε.

The control volumes method is applied for the discretization of the PDEs of the mathematical model. This method combines the main advantages of the finite elements method and finite differences method (Stankov 1998), namely:

- Freedom of choice of the geometry in grid generation, typical for the finite elements method.
- Freedom in the determination of the discrete values of the dependent variables, typical for the finite differences method.

The integration over the control volume distinguishes the control volumes method from all other methods in CFD, as it satisfies the laws of conservation. The integral laws for *mass, momentum, and energy conservation*, which are set in the output differential equations, are set in the method of control volumes as well. The result reflects the conservation of the fluid characteristics for each cell, as the conservation laws are applied to each control volume. This clear correlation between numerical algorithm and the physical phenomenon is one of the main advantages of the method of control volumes.

Another advantage of the method is the geometry of the control volumes—they may have different shapes. This and other opportunities associated with the discretization make the method very common in modeling of practical engineering problems, as is the case in this study. As a result of the discretization, the nonlinear partial differential equations transform into a system of linear algebraic equations, whose solution is the major task of the solvers.

However, iterations are necessary to solve even one of the differential equations. Therefore, the system of algebraic equations is solved through a certain algorithm that uses an iterative approach. Moreover, the relation between the particular PDEs is taken into account, that is, the relation between the transport equations and the continuity equation, as well as the relation between the transport equations and the equations of the turbulent model.

The most popular algorithm that uses an iterative approach to solve systems of algebraic equations is the SIMPLE (semi-implicit procedure for pressure-linked equations) algorithm (Wendt 1992). The reason for the wide-spread use of such an algorithm is the necessity of determining the pressure distribution along with the other parameters of the flow, as in the general case of simulation of an incompressible fluid the pressure gradient across the computational domain is unknown.

14.4 Theoretical Background of the Modeling of Woven Macrostructures

Theoretically, the complex hierarchical structure of the woven textiles requires complex hierarchical approach to simulate the processes of heat and mass transfer through it (Angelova 2010b). Practically, there are, in the literature, different approaches to model the woven macrostructure, which are as follows:

- By geometric description of a *unit cell*, built of two warp and two weft threads
- By geometric description of one repeat of the weave
- By presenting the woven macrostructure as a perforated plate or a net of an infinite number of openings
- By describing the woven macrostructure as a continuum of fibers
- By presenting the fibers in the macrostructure as bodies flowed by fluid

When simulating the air permeability of a woven structure, the first three ways are reduced, in fact, to the simulation of flow through openings. The fourth way corresponds to the simulation of flow through porous media, and the fifth—to simulation of a flow around a body.

In this part of the study, an original approach for simulation of the flow through a woven structure is presented (Angelova et al. 2011). The approach is based on the theory of jet systems and systematic experimental investigations on this topic (Stankov et al. 1994; Lozanova et al. 1997; Lozanova and Stankov 1998; Stankov 1998; Stankov and Simova 2010).

14.4.1 Jet Systems

The jet system results from the interaction of free parallel turbulent jets, located in different geometrical order in the output plane (Stankov 1998). The morphology of the pores in the woven macrostructure is similar to a jet system, through which a flow can pass if a pressure gradient exist between the two sides of the textile barrier.

The resulting flow in the case of jet systems is a complex 3D flow. It is formed upon the interaction of single jets, which flow out of openings with a circular, square, or rectangular cross section. Single jets can be arranged in schemes, defined in Stankov (1998) and visualized in Figure 14.1:

- *In-line-ordered jet systems:* This is the simplest arrangement of a jet system (Figure 14.1a). Every five adjacent jets are sufficiently *representative* for a system of an infinite number of jets.

- *In-corridor-ordered jet systems:* Here a system of an infinite number of jets is described by a single *element* of nine jets (Figure 14.1b).

- *Chess-board-ordered jet systems:* Each *element* of seven jets is *representative* for the whole system of an infinite number of jets (Figure 14.1c).

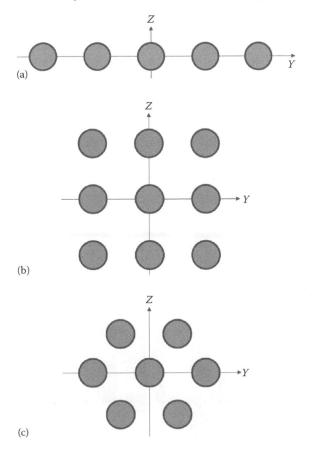

FIGURE 14.1
(a) In-line-ordered jet system; (b) in-corridor-ordered jet system; and (c) chess-board-ordered jet system.

Stankov (1998) proved that a single jet can be representative for all other jets in the jet system if it is surrounded by other identical jets and the whole *pattern* is investigated either experimentally or numerically. In the case of the corridor-ordered jets, the minimum number of jets is eight: the eight surrounding jets interfere with the central jet and play the role of *boundary conditions* that influence the development of the central jet.

14.4.2 Approximation of a Woven Structure to a Jet System

The macrostructure of the woven textiles has a strictly defined geometry of alternating warp and weft threads and air (pores) between them. Therefore, the flow that passes in through-thickness direction can be approximated to a system of jets and specifically to a corridor-ordered jet system (Angelova et al. 2011, 2013). This approach for approximation of the woven macrostructure is similar to that used in the literature, when textiles are presented as nets, perforated plates, a *cell unit* of two warp and two weft threads, and a repeat of the weave pattern. The difference is that the jet systems approach allows each pore between the threads (yarns or filaments) to be treated as a representative for the macrostructure provided that the remaining eight pores around it play the role of boundary conditions. Figure 14.2 visualizes this approximation.

The basic parameters of a jet system are as follows:

- The steps S_y and S_z between the single free jets, along the coordinates Y and Z, respectively.
- The geometry of the output cross section.
- The equivalent diameter (in case of a circular cross section) or equivalent side (in case of a rectangular cross section).

The approximation approaches use the following parameters of the woven macrostructure (Angelova et al. 2013):

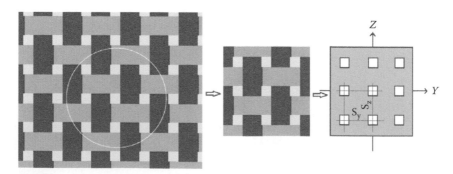

FIGURE 14.2
Approximation of a woven structure to an in-corridor-ordered jet system.

- Mean diameter of the warp threads d_{wa}.
- Mean diameter of the weft threads d_{wf}.
- Equivalent diameter or side of the pore, when the mean pore area S_{av} is measured or calculated.

On this basis, the main parameters of the jet system are determined (Figure 14.2):

$$S_y = 2b_0 + d_{wa} \tag{14.6}$$

$$S_z = 2b_0 + d_{wf} \tag{14.7}$$

where:
   b_0 is the side length of the square pore

The study of Stankov (1998) has proved that a single jet is representative for the whole system of corridor-ordered jets, if it is surrounded by eight other jets. This allows the numerical task to be facilitated in terms of creating the virtual model, the grid generation, and the calculation itself (reduction of the CPU time). The resulting flow, however, is undoubtedly 3D.

   The performed numerical study on the air permeability of woven macrostructures in through-thickness direction involved two different numerical tasks:

- *Parametric study* to determine the conditions for the numerical simulation.
- *Model study* on the air permeability of real woven macrostructures.

The numerical procedure for each task and the results from the simulations are presented separately.

14.5 Numerical Procedure for the Parametric Study

14.5.1 Computational Domain

The selected approach for modeling of the woven macrostructure allowed to limit the size of the computational domain, seeking symmetries or other characteristic parameters. In this case, the boundary conditions on the sides of the computational domain were set as symmetrical (Simova et al. 2009). The boundary conditions at the inlet and outlet of the computational domain had to be set at a sufficient distance from the textile layer (away from the presence of gradients) and to provide enough space/time for the flow development downstream after the textile barrier. Similar approach was demonstrated in the studies of Nazarboland et al. (2007), Saldaeva (2010), and Duru

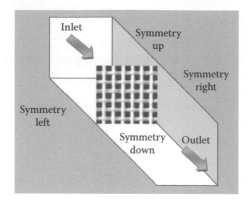

FIGURE 14.3
Scheme of the computational domain and part of the boundary conditions.

Cimilli et al. (2012). Figure 14.3 presents the basic scheme of the computational domain and part of the relevant boundary conditions.

The woven macrostructure was built from elements with dimensions of $0.225 \times 1.365 \times 1.365$ mm^3, placed in the volume of computational domain, as shown in Figure 14.3. The distance from the beginning of the computational domain was set to 4 mm, and the length of the domain after the model of the woven macrostructure was 8 mm. These distances were determined on the basis of preliminary tests so as to ensure the development of the downstream flow (Angelova et al. 2011).

14.5.2 Selection of the Woven Macrostructure

The selected woven macrostructure was sample 5 from Chapter 7, a cotton fabric for bedding and clothing (see Tables 7.3 and 7.4, Chapter 7). The fabric weight of the sample was 138 g/m^2, woven in a plain weave from threads with 25 tex linear density. The warp density of the sample was 270 threads/dm, the weft density was 266 threads/dm, and the pore area was 0.024 mm^2.

It was discussed in Section II the need to assess the approximation of the shape of the pore to a circular or square cross section, in the presence of absolutely free geometry of the pores in the macrostructures, woven from staple fiber yarns. Such an assessment could be made only by means of numerical simulation. Therefore, one of the aims of the parametric study was to evaluate the influence of the shape of the pore cross section on the flow development after the textile *barrier* and on the air permeability. To this end, the average pore size of the selected macrostructure was approximated to

- Pore with a circular cross section with an equivalent diameter of $d_{eqv} = 0.175$ mm.
- Pore with a square cross section with an equivalent side of the square of $a_{eqv} = 0.155$ mm.

14.5.3 Selection of the Jet System

The numerical study gave the possibility to check the experimental conclusions of Stankov (1998) for the representativeness of the central jet of a corridor-order jet system, if the central jet is surrounded by eight equal jets (scheme 3 × 3 jets).

In the parametric study, the selected woven macrostructure was simulated as three types of corridor-ordered jet systems (concerning the number of jets in the system):

- System of 3 × 3 jets, that is, the simulated woven macrostructure involves 4 × 4 threads.
- System of 5 × 5 jets, that is, the simulated woven macrostructure involves 6 × 6 threads.
- System of 10 × 10 jets, that is, the simulated woven macrostructure involves 11 × 11 threads.

14.5.4 Selection of a Grid

The computational grid is a determining factor for the computational study and preconditions the quality of the numerical results. This is especially true in the case of simulation of air permeability through woven macrostructure in through-thickness direction, since the computational grid must comply with the very small size of the threads, the pores between them, and the thickness (Simova et al. 2009).

The physical nature of the computational task allowed to use both structured (Figure 14.4a) and unstructured grid (Figure 14.4b).

The requirement for a rectangular topology of the control volumes in the generation of structured grids creates limitations, particularly if elements of circular cross section are involved (as in the case of simulation of pores with circular cross section of the woven macrostructure). The unstructured grids in turn allow the use of a wide range of different shapes of the control

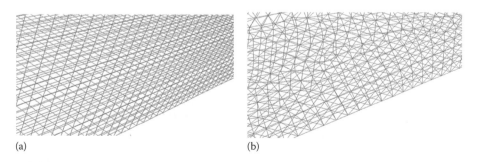

(a)   (b)

FIGURE 14.4
(a) Structured grid of the computational domain and (b) unstructured grid of the computational domain.

volumes: tetrahedron, hexagon, pyramid, and so on. Both structured and unstructured grids allow local refinement of the grid in zones where the larger gradients require more precise calculation.

Hybrid grids can also be applied for meshing the computational domain. In such a case, parallelepiped control volumes are generated in the near-wall regions (to account for the viscous effects in the boundary layer), and control volumes with a hexagonal or other shape are used in the rest of the volume of the computational domain (Simova et al. 2009).

The numerical experiment had as an aim to test the different types of grids, available in the FLUENT preprocessor GAMBIT. Three types of grids were used for simulating the cases with *circular cross section of the pores* (Angelova et al. 2011):

- Hybrid grid (control volumes in the shape of parallelepipeds and tetrahedrons), which was additionally refined.
- Structured grid (control volumes in the shape of parallelepipeds), which was additionally refined.
- Unstructured grid (control volumes in the shape of tetrahedrons).

Two types of grids were used for the simulation of the cases with *square cross section of the pores*:

- Structured grid (control volumes in the shape of parallelepipeds), which was additionally refined.
- Unstructured grid (control volumes in the shape of tetrahedrons).

The data for the generated grids in the parametric study are summarized in Table 14.1 for the cases with circular cross section of the pore and Table 14.2 for the cases with square cross section of the pore. The number of cells, faces, and nodes for each type of grid are indicated.

TABLE 14.1

Grid Data: Cases with Circular Cross Section of the Pore

Jet System	Pore Cross Section	Grid Type	3D Cells	Faces	Nodes
3 × 3	●	Hybrid	440,275	908,368	93,402
		Structured	206,412	632,364	219,683
		Structured with local refinement	1,101,520	3,344,520	1,141,650
		Unstructured	622,438	1,281,171	122,186
5 × 5	●	Hybrid	440,113	929,333	105,232
10 × 10	●	Hybrid	939,903	2,027,623	248,240

TABLE 14.2

Grid Data: Cases with Square Cross Section of the Pore

Jet System	Pore Cross Section	Grid Type	3D Cells	Faces	Nodes
3 × 3	■	Structured	580,796	1,770,243	608,788
		Structured with local refinement	531,360	1,625,724	562,965
		Unstructured	165,342	349335	37,069
5 × 5	■	Structured	768,222	2,339,877	803,392
10 × 10	■	Unstructured	1,457,223	3,070,333	335,003

14.5.5 Turbulent Model

A standard $k - \varepsilon$ turbulent model was used, so that two additional transport equations for the kinetic energy k and its dissipation rate ε were added to the system of RANS partial differential equations:

$$\frac{\partial}{\partial x}\left[\rho u k - \left(\mu + \frac{\mu_t}{\sigma_k}\right)\frac{\partial k}{\partial x}\right] + \frac{\partial}{\partial y}\left[\rho v k - \left(\mu + \frac{\mu_t}{\sigma_k}\right)\frac{\partial k}{\partial y}\right]$$

$$+ \frac{\partial}{\partial z}\left[\rho w k - \left(\mu + \frac{\mu_t}{\sigma_k}\right)\frac{\partial k}{\partial z}\right] = P - \rho\varepsilon \tag{14.8}$$

$$\frac{\partial}{\partial x}\left[\rho u \varepsilon - \left(\mu + \frac{\mu_t}{\sigma_\varepsilon}\right)\frac{\partial \varepsilon}{\partial x}\right] + \frac{\partial}{\partial y}\left[\rho v \varepsilon - \left(\mu + \frac{\mu_t}{\sigma_\varepsilon}\right)\frac{\partial \varepsilon}{\partial y}\right]$$

$$+ \frac{\partial}{\partial z}\left[\rho w \varepsilon - \left(\mu + \frac{\mu_t}{\sigma_\varepsilon}\right)\frac{\partial \varepsilon}{\partial z}\right] = \frac{\varepsilon}{k}\left(c_1 G - c_2 \rho\varepsilon\right) \tag{14.9}$$

The generation G is described by the expression

$$G = \mu_t\left\{2\left[\left(\frac{\partial u}{\partial x}\right)^2 + \left(\frac{\partial v}{\partial u}\right)^2 + \left(\frac{\partial w}{\partial z}\right)^2\right] + \left(\frac{\partial u}{\partial y} + \frac{\partial v}{\partial x}\right)^2 + \left(\frac{\partial v}{\partial z} + \frac{\partial w}{\partial y}\right)^2\right.$$

$$\left. + \left(\frac{\partial w}{\partial x} + \frac{\partial u}{\partial z}\right)^2\right\} \tag{14.10}$$

The effective viscosity is determined by the sum

$$\mu_{eff} = \mu + \mu_t \tag{14.11}$$

where for the turbulent viscosity μ_t is valid:

$$\mu_t = c_\mu \rho \frac{k^2}{\varepsilon} \tag{14.12}$$

The model constants are $c_\mu = 0.09$, $\sigma_k = 1$, $\sigma_\varepsilon = 1.3$, $C_1 = 1.44$, and $C_2 = 1.92$.

14.5.6 Initial and Boundary Conditions

Symmetric boundary conditions were set for the walls. The following values of the initial flow velocity were studied:

- 0.4, 2, and 10 m/s for the cases with square cross section of the pore
- 0.4 and 2 m/s for the cases with circular cross section of the pore

The initial velocity of 0.4 m/s corresponded to the flow movement of the air in the indoor environment. The value of 2 m/s corresponded to the flow velocity in special cases of enclosures, that is, fridges or next to fans, or was typical for the outdoor environment, the same as the value of 10 m/s (windy weather).

14.5.7 Convergence Conditions

In most cases, one and the same condition for convergence was used: the absolute value of the residuals was set to 1×10^{-4}. In some cases, however, it was necessary to use a more restrictive criterion for convergence: 1×10^{-6} for the continuity equation and 1×10^{-5} for the other variables.

14.5.8 Additional Conditions

The following simplifications of the physical task were made due to the complexity of the woven macrostructure and the approximation of the air flow through the pores as a jet system (Angelova et al. 2011):

- The linear density of the threads did not change in length.
- The yarns interlacing (weave pattern) affected the thickness of the macrostructure, which was determined experimentally or calculated. The thickness determined the length of the pore.
- The size of the pores (or the equivalent diameter/equivalent side) was set to be equal for all pores within the virtual model.

14.5.9 Summary of the Investigated Cases

All 37 cases in the parametric study are summarized in Table 14.3. Figure 14.5 presents the computational domain for case 1: structured grid for 3 × 3 jet system with square cross section of the pores. Figure 14.6 shows the local

TABLE 14.3

Simulated Cases in the Parametric Study

Case	Jet System	Pore Cross Section	Initial Velocity V_{in}, m/s	Grid Type	Convergence Criterion
1	3×3	■	0.4	Structured	Standard
2	3×3	■	0.4	Unstructured with local refinement	Standard
3	3×3	■	0.4	Unstructured	Standard
4	3×3	■	2	Structured	Standard
5	3×3	■	2	Unstructured with local refinement	Standard
6	3×3	■	2	Unstructured	Standard
7	3×3	■	2	Unstructured	Modified
8	3×3	■	10	Structured	Standard
9	3×3	■	10	Unstructured with local refinement	Standard
10	3×3	■	10	Unstructured	Standard
11	3×3	■	10	Unstructured	Modified
12	3×3	■	10	Structured	Modified
13	5×5	■	0.4	Structured	Standard
14	5×5	■	2	Structured	Standard
15	5×5	■	10	Structured	Standard
16	10×10	■	0.4	Unstructured	Standard
17	10×10	■	0.4	Unstructured	Modified
18	10×10	■	2	Unstructured	Standard
19	10×10	■	2	Unstructured	Modified
20	10×10	■	10	Unstructured	Standard
21	10×10	■	10	Unstructured	Modified
22	3×3	●	0.4	Hybrid	Standard
23	3×3	●	0.4	Structured	Standard
24	3×3	●	0.4	Unstructured with local refinement	Standard
25	3×3	●	0.4	Unstructured	Standard
26	3×3	●	2	Hybrid	Standard
27	3×3	●	2	Structured	Standard
28	3×3	●	2	Unstructured with local refinement	Standard
29	3×3	●	2	Unstructured	Standard
30	5×5	●	0.4	Hybrid	Standard
31	5×5	●	0.4	Hybrid	Modified
32	5×5	●	2	Hybrid	Standard
33	5×5	●	2	Hybrid	Modified
34	10×10	●	0.4	Hybrid	Standard
35	10×10	●	0.4	Hybrid	Modified
36	10×10	●	2	Hybrid	Standard
37	10×10	●	2	Hybrid	Modified

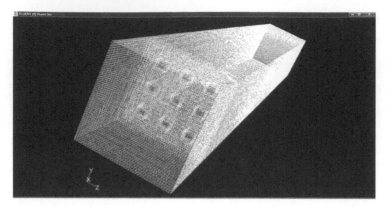

FIGURE 14.5
Computational domain, case 1: 3 × 3 jet system, square pores, and structured grid.

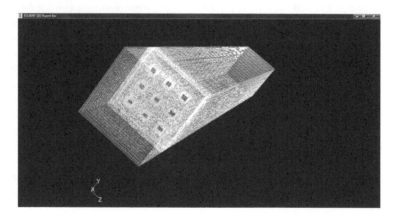

FIGURE 14.6
Computational domain, case 2: 3 × 3 jet system, square pores, and structured grid with local refinement.

refinement of the computational grid in the zone near the woven macro-structure (case 2).

Figure 14.7 illustrates the computational grid for case 15: structured grid for a 5 × 5 jet system with square cross section of the pores. Unstructured grid for a 10 × 10 jets system with square pores (case 18) is shown in Figure 14.8. The computational domain for case 22 is presented in Figure 14.9: a 3 × 3 jet system with circular cross section of the pores, hybrid grid.

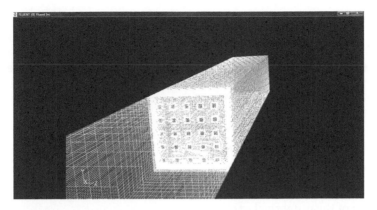

FIGURE 14.7
Computational domain, case 15: 5 × 5 jet system, square pores, and structured grid.

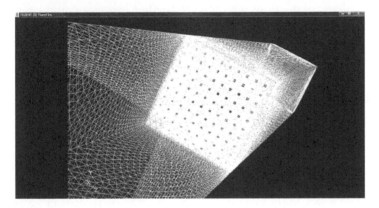

FIGURE 14.8
Computational domain, case 18: 10 × 10 jet system, square pores, and unstructured grid.

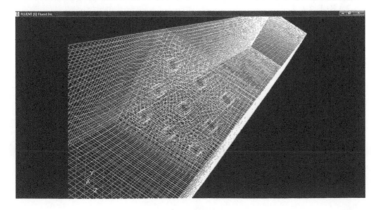

FIGURE 14.9
Computational domain, case 22: 3 × 3 jet system, circular pores, and hybrid grid.

14.6 Numerical Results from the Parametric Study

Exemplary numerical results from the parametric study are shown hereinafter. They permit to evaluate the influence of the studied factors on the numerical simulation of the air permeability of woven structure by using the jet system approach:

- Number of jets in the system
- Cross-sectional shape of the pore
- Initial velocity of the air flow
- Type of the numerical grid

For each of the simulated cases, numerical database was obtained for the local parameters of the flow, which were further processed and presented in the following charts: velocity profiles, velocity contours, axial velocity decay, and so on. The numerical results were presented in a series of cross sections of the flow after the woven macrostructure, listed in Table 14.4.

14.6.1 Effect of the Number of Jets in the System

Figures 14.10 through 14.12 visualize the dimensionless velocity profiles for jet systems, consisted of 3×3, 5×5, and 10×10 jets, respectively, with a square cross section of the pore and 2 m/s initial velocity of the flow (cases 4, 14, and 18). The velocity profiles showed qualitative and quantitative conformity, regardless of the number of the jets in the system, following the logic of development of jet flows. In the closest to the textile macrostructure cross section of the flow (lin a), the maximum velocity $U/U_0 = 1$ was equal for the three cases.

TABLE 14.4

Cross Sections of the Flow after the Woven Macrostructure

Cross Section	Coordinate, mm	Distance from the Macrostructure, mm
lin a	0.1125	0
lin c	0.3125	0.2
lin e	0.5125	0.4
lin g	0.7125	0.6
lin i	0.9125	0.8
lin k	1.1125	1
lin m	1.5125	1.4
lin o	1.9125	1.8
lin q	2.3125	2.2
lin s	2.7125	2.6

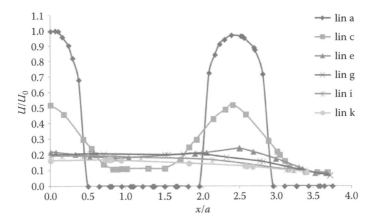

FIGURE 14.10
Velocity profiles, case 4: 3 × 3 jet system, square pores, 2 m/s.

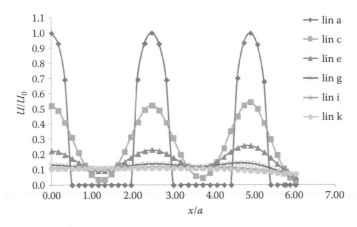

FIGURE 14.11
Velocity profiles, case 14: 5 × 5 jet system, square pores, 2 m/s.

In the next cross section, distancing at 0.2 mm (lin c), the maximum velocity along the axis of the jet was again equal for the three cases: $U/U_0 = 0.52$.

The comparison of the velocity contours for cases 4 and 14 showed again similarity in the development of the jet flow in the cross sections near the woven macrostructure (Figures 14.13 and 14.14). The impact of the number of jets in the system was noticeable with moving away from the textile *barrier*, actually in the zone of transformation of the jets currents in a uniform flow.

In fact, the nature of the physical task, associated with thermophysiological comfort and the air permeability, requires an assessment of the coincidence of the flow development from different types of jet systems in the cross

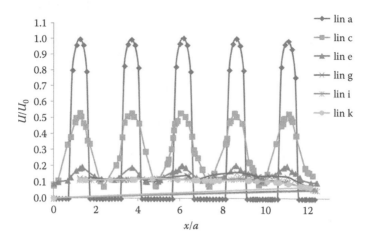

FIGURE 14.12
Velocity profiles, case 18: 10 × 10 jet system, square pores, 2 m/s.

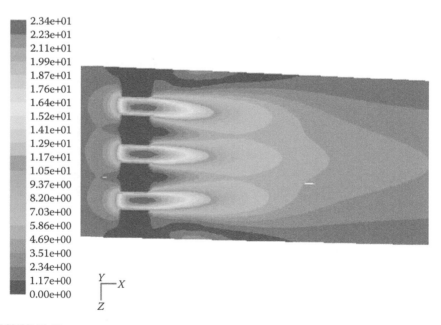

FIGURE 14.13
Velocity contours, m/s, case 4: cross section $y = 0$, 3 × 3 jet system, square pores, 2 m/s.

sections immediately after and near the woven macrostructure. In cases, however, when the flow development away from the macrostructure is of importance, as in the case of precision woven macrostructures (Sefar 2008), the results require careful analysis of the interchangeability of the simulated cases, that is, types of jet systems.

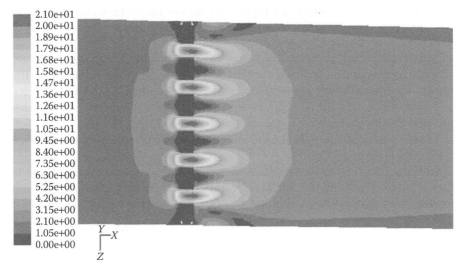

FIGURE 14.14
Velocity contours, m/s, case 14: cross section $y = 0$, 5×5 jet system, square pores, 2 m/s.

The main conclusion from this part of the numerical study is that *3 × 3 jet system is representative for a system of countless in-corridor-ordered jets.* The numerical results proved that the simulation of air permeability of woven macrostructures through a system of in-corridor-ordered 3 × 3 jets leaded to reliable quantitative and qualitative results. Therefore, the creation of a virtual model with higher number of threads/pores involved is meaningless. This reduces both the efforts of the operator and time for the numerical simulation.

Based on the results obtained within the parametric study, a system of 3 × 3 in-corridor-ordered jets was selected for the model study of the air permeability of woven macrostructures.

14.6.2 Effect of the Shape of the Pore Cross Section

The shape of the cross section of the pores in the woven macrostructure depends largely on the mesostructure. Only in the case of weaving with monofilament or strips (similar to the macrostructures for packaging, investigated in Chapter 10), the shape of the pore cross section is close to a rectangle. In the case of weaving with staple fiber yarns or polyfilaments (including twisted polyfilaments), the *corners* of the pores are rounded due to the friction between the threads in the process of shed forming or beating the weft thread. These make part of the fibers or filaments of the mesostructure to separate from the core. Thus, the shape of the pores between the threads in the macrostructure is closer to a circle than to a rectangle.

The simulation of the process of air permeability of the woven macrostructure by means of CFD allows, in contrast to the experimental studies, to evaluate the impact of the shape of the pore for theoretical calculations of the air permeability, that is, by using regression models or the equation of Hagen–Poiseuille (see Chapter 13).

Figures 14.15 and 14.16 visualize the velocity profiles for cases 1 and 23, which differ only in terms of the shape of the pore. The analysis of the graphs for the different cross sections of the flow after the woven macro-structure, alongside the X-axis, showed that there were neither qualitative nor quantitative differences between the flow development in the two cases. The same was confirmed by the velocity contours for the same cases, shown in Figures 14.17 and 14.18.

Studying the axial velocity decay alongside the computational domain is also a source of information for the development of the flow. The comparison between the cases with circular and square cross section of the pore showed

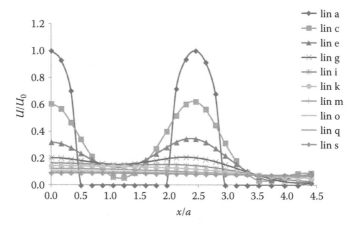

FIGURE 14.15
Velocity profiles, case 1: 3 × 3 jet system, square pores, 0.4 m/s.

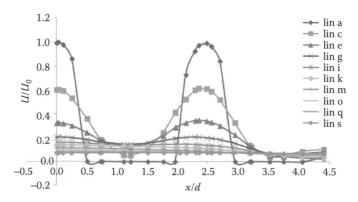

FIGURE 14.16
Velocity profiles, case 23: 3 × 3 jet system, circular pores, 0.4 m/s.

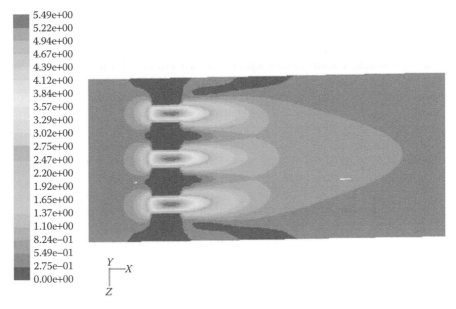

FIGURE 14.17
Velocity contours, m/s, case 1: cross section $y = 0$, 3×3 jet system, square pores, 0.4 m/s.

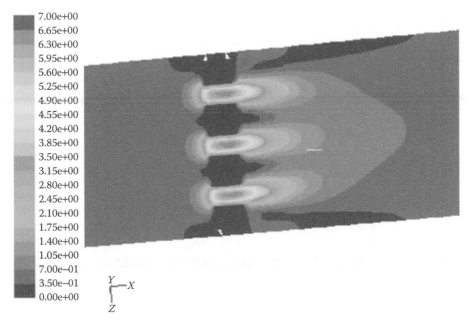

FIGURE 14.18
Velocity contours, m/s, case 1: cross section $y = 0$, 3×3 jet system, circular pores, 0.4 m/s.

that there was no influence of the shape of the pore on the flow development after the textile macrostructure.

The main conclusion from the numerical results is that the approximation of the pore shape to a circular or square cross section is not important for the theoretical calculations and modeling studies, if the simulated pore area corresponds to the real pore area. This can be explained with the fast transformation of the jet, issuing from an opening with a square cross section, into a jet with a round cross section, very near to the outlet. The practical result of the simulation of the air permeability of a woven structure in through-thickness direction is *that the pore approximation to a particular geometric shape is unrestricted* and will depend on the particular task, the selection of the type of the computational grid, and, last but not least, on the experience of the operator in the building of the virtual model and grid generation.

Based on the results obtained within the parametric study, a 3 × 3 jet system with a square cross section of the pores was applied in the model study on the air permeability of woven macrostructures.

14.6.3 Effect of the Initial Flow Velocity

Figures 14.19 through 14.21 present the velocity profiles for 5 × 5 jet systems with a square cross section of the pores and structured computational grid, with different values of the initial flow velocity (cases 13–15). The assessment of the impact of the initial velocity on the flow development aimed to test the reaction of the model, moreover that the selected velocity values corresponded to real conditions. Along with this, each successive value was five times higher than the previous one, namely, 0.4, 2, and 10 m/s.

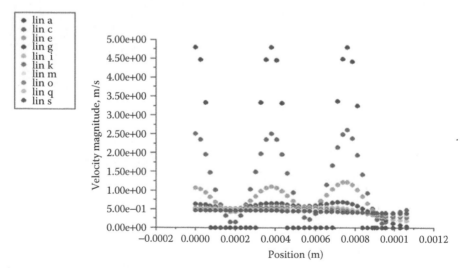

FIGURE 14.19
Velocity profiles, case 13: 5 × 5 jet system, square pores, structured grid, 0.4 m/s.

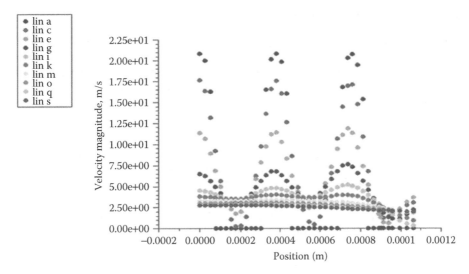

FIGURE 14.20
Velocity profiles, case 14: 5 × 5 jet system, square pores, structured grid, 2 m/s.

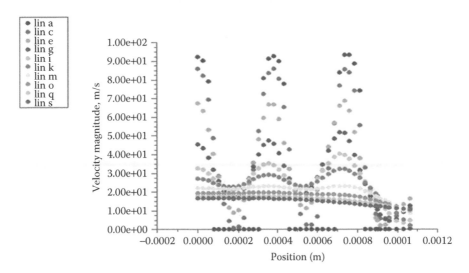

FIGURE 14.21
Velocity profiles, case 15: 5 × 5 jet system, square pores, structured grid, 10 m/s.

The analysis of the numerical results showed that the velocity profiles reasonably reflected the interaction between the jets in the system and the development of the flow downstream after the textile macrostructure for all three cases. Certainly, the velocity of the resulting flow increased with the increment of the initial flow velocity, set as an initial condition of the simulation.

14.6.4 Effect of the Numerical Grid

The type of the numerical grid plays significant role for the exactness and the stability of the simulation. Figure 14.22 presents the numerical results for the axial velocity decay for cases 1–3, where the effect of three types of grids was tested: structured, structured with a local refinement, and unstructured. The other parameters of the simulation were set to be equal: 3 × 3 jet system with a square cross section of the pore and 0.4 m/s initial velocity of the flow.

The results clearly showed that there was no significant difference between the simulated cases with structured network without and with local refinement. This is probably due to the relatively nearest number of control volumes (Table 14.2). The unstructured grid, however, gave significant deviations in the results for the $X/a = 1/5$ cross sections. These were precisely the cross sections in which the results of the air transfer had the greatest significance in terms of the physical task: air permeability of woven macrostructures related to thermophysiological comfort. The deviations could be related either with the different geometry of the control volumes (tetrahedrons) or with the significantly lower number of control volumes in the generation of the grid: approximately 165,000 in the unstructured grid against over 500,000 in the structured one.

Similar analyzes can be made for the results in Figure 14.23, which visualizes the axial velocity decay for the cases 26–29: 3 × 3 jet systems with a circular cross section of the pores and initial velocity of 2 m/s.

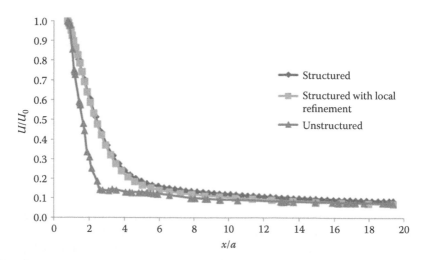

FIGURE 14.22
Axial velocity decay: comparison between three types of grids, 3 × 3 jet system, square pore, 0.4 m/s.

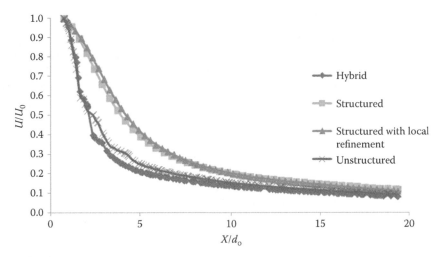

FIGURE 14.23
Axial velocity decay: comparison between four types of grids, 3 × 3 jet system, circular pore, 2 m/s.

Structured grid without and with local refinement gave again similar results, despite the difference in the number of the control volumes, which was five times higher for the grid with a local refinement (Table 14.1). Therefore, the local refinement in the case would only increase the time of calculation.

The hybrid grid influenced the numerical results on the same way as the unstructured grid. Both grids, however, gave different values for the velocity in the cross sections $X/d = 1/12$, compared to the structured grid with and without local refinement.

As a whole, it was found that the numerical results for the axial velocity decay of the resulting flow from a jet system, obtained through a *structured grid*—with and without local refinement—described more correctly the decay of the resulting flow, compared to the other types of grids used. Therefore, the use of structured grid was preferred in the model study of the air permeability of woven macrostructures.

14.7 Numerical Procedure for the Model Study

The numerical procedure for the *model study* on the air permeability of real woven macrostructures was essentially determined by the results from the *parametric study*, which helped the basic parameters of the numerical simulation to be set (Angelova et al. 2013).

14.7.1 Computational Domain

The woven macrostructure was placed in the volume of the computational domain, as already described in the parametric study. It was built from elements with dimensions $0.45 \times 2.232 \times 2.232 \text{ mm}^3$. The distance from the inlet of the computational domain to the woven macrostructure was 4 mm, and from the woven macrostructure to the outlet was 8 mm.

14.7.2 Selection of the Woven Macrostructure

Real woven macrostructures for clothing and bedding, investigated experimentally in Chapter 7, were selected: samples 2, 4, 6, 7, and 10. The five samples differed from each other in geometrical, structural, and mass parameters, as shown in Table 14.5. The equivalent side a_{eqv} of their pores with square cross section was determined following the described methodology in Chapter 7.

14.7.3 Selection of the Jet System

All woven macrostructures were approximated to a 3×3 in-corridor-ordered jet system, that is, 4×4 threads of each sample were modeled. Square cross section of the pores was simulated.

14.7.4 Selection of a Grid

A structured grid was generated for each of the samples with data, summarized in Table 14.6.

TABLE 14.5

Data of the Simulated Woven Macrostructures

Sample	$P_{wa'}$ threads/ dm	$P_{wf'}$ threads/ dm	$Tt_{wa'}$ tex	$Tt_{wf'}$ tex	Thickness, mm	Weave	Fabric Weight, g/m²	$a_{eqv'}$ mm
2	176	124	28	28	0.45	Plain	89	0.505
4	284	294	20	20	0.39	Plain	133	0.199
6	234	254	30	30	0.47	Plain	157	0.214
7	383	338	28	28	0.68	Twill 2/1Z	184	0.250
10	386	286	30	36	0.71	Twill 3/1Z	202	0.156

TABLE 14.6

Grid Data for the Model Study

Sample	Size of the Control Volume, mm	3D Cells	Faces	Nodes
2	0.0621889	237,168	729,432	255,192
4	0.0649915	127,575	394,848	140,020
6	0.0584984	214,533	660,570	231,620
7	0.0600331	310,400	952,240	331,539
10	0.0504057	608,634	1,858,923	641,772

14.7.5 Turbulent Model

Three turbulent models were tested in the model study: $k - \varepsilon$ and $k - \omega$ from the group of turbulent models with eddy viscosity (EVM) and RSM, and turbulent model with Reynolds stresses (Angelova et al. 2013):

- $k - \varepsilon$ *turbulent model*: Two additional transport equations are added to the RANS system of PDEs: one for the kinetic energy k and one for its dissipation rate ε.

- $k - \omega$ *turbulent model*: Two additional transport equations are added as well, but instead an equation for the dissipation of the kinetic energy ε, and equation for the specific dissipation ω is added.

- *RSM turbulent model*: It is a higher class turbulent model, from the group of second-order closures. It adds additional transport equations for the Reynolds stresses so as they can be directly calculated, if necessary for a more detailed analysis of the flow (i.e., analysis of the flow immediately after the textile *barrier*).

14.7.6 Initial and Boundary Conditions

Symmetric boundary conditions were set on the walls of the computational domain. A pressure difference of 100 Pa was set between the inlet and the outlet, so as to be similar to the conditions for experimental measurement of the air permeability (see Chapter 5).

14.7.7 Convergence Conditions

The residuals for the continuity equation were set to 1×10^{-6}, and for the rest of the values: 1×10^{-5}.

14.7.8 Additional Conditions

The same conditions for simplification of the physical task were kept, as described in the detail for the parametric study.

14.7.9 Summary of the Investigated Cases

Table 14.7 summarizes the cases, numerically investigated within the model study. Fifteen cases all together were simulated.

The computational domains for sample 2 (cases 1–3) and sample 4 (cases 4–6) are shown in Figures 14.24 and 14.25, respectively.

TABLE 14.7

Simulated Cases in the Model Study

Case	Sample	Jet System	Pore Cross Section	Turbulent Model
1	2	3×3	■	$k - \varepsilon$
2	2	3×3	■	$k - \omega$
3	2	3×3	■	RSM
4	4	3×3	■	$k - \varepsilon$
5	4	3×3	■	$k - \omega$
6	4	3×3	■	RSM
7	6	3×3	■	$k - \varepsilon$
8	6	3×3	■	$k - \omega$
9	6	3×3	■	RSM
10	7	3×3	■	$k - \varepsilon$
11	7	3×3	■	$k - \omega$
12	7	3×3	■	RSM
13	10	3×3	■	$k - \varepsilon$
14	10	3×3	■	$k - \omega$
15	10	3×3	■	RSM

 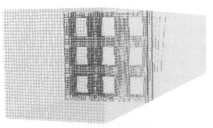

Y
$\llcorner Z$
X

FIGURE 14.24
Computational grid, sample 2 (cases 1–3).

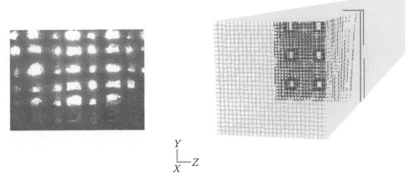

FIGURE 14.25
Computational grid, sample 4 (cases 4–6).

14.8 Numerical Results from the Model Study

14.8.1 Effect of the Turbulent Model

Figures 14.26 through 14.28 visualize the velocity field of the air flow, passing through sample 7, cases 10–12 (Table 14.7), when applying different turbulent models. The results are exemplary, but they showed a tendency, typical for all samples: $k - \varepsilon$ (Figure 14.26) and RSM (Figure 14.27) turbulent models gave similar results as they concerned the flow development and the velocity.

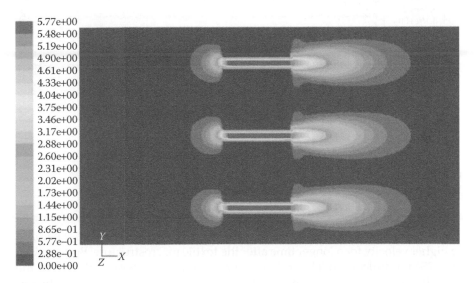

FIGURE 14.26
Velocity contours, m/s, sample 7, case 10: $k - \varepsilon$ turbulent model.

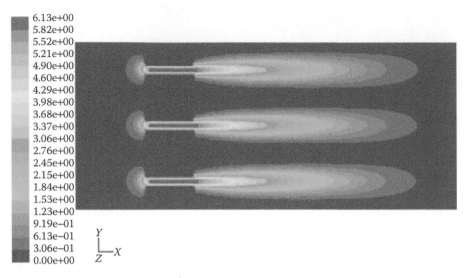

FIGURE 14.27
Velocity contours, m/s, sample 7, case 11: $k - \omega$ turbulent model.

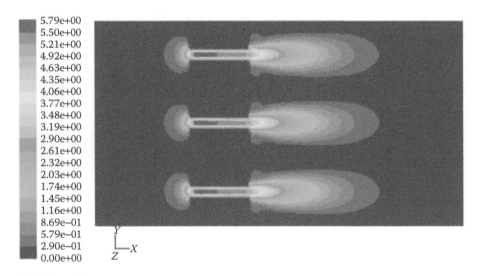

FIGURE 14.28
Velocity contours, m/s, sample 7, case 12: RSM turbulent model.

The application of $k - \omega$ (Figure 14.28) turbulent model provoked the appearance of higher velocity for a longer time after the textile macrostructure. Besides, $k - \varepsilon$ and RSM turbulent models led to the simulation of shorter initial zone of the resulting flow (defined in Stankov 1998) in comparison with the $k - \omega$ turbulent model.

To quantify the effect of the turbulent model on the simulation of the through-thickness air permeability of the woven macrostructure, the velocity profiles after the textile sample had to be analyzed. To this end, the same cases of the simulation of sample 7 were considered, so as to make a comparison with the visualization of the velocity contours in Figures 14.26 through 14.28. The velocity profiles in cross sections $X = 0.8, 1, 2, 3, 4, 5$ mm after the woven macrostructure were presented.

The analysis of the figures showed that the maximum velocity of the flow after the textile macrostructure was achieved with the application of $k - \omega$ turbulent model: 5.81 m/s (Figure 14.30). This value for the simulation with $k - \varepsilon$ turbulent model was 5.03 m/s (Figure 14.29), and the use of RSM turbulent model led to a maximum velocity of 4.52 m/s (Figure 14.31).

There was a considerable difference between the velocity decay of the single jets in the jet system. For example, the velocity of the jet flow for the cross section, distanced at 1 mm after the woven macrostructure, was 3.65 m/s for the simulation with $k - \varepsilon$ turbulent model (Figure 14.29), 5.32 m/s for the simulation with $k - \omega$ turbulent model (Figure 14.30), and 3.4 m/s for the simulation with RSM turbulent model (Figure 14.31).

The analogous study of the cross section situated at 3 mm after the woven macrostructure showed that the flow velocity in the simulation with $k - \varepsilon$ and RSM turbulent models was close to 0, while in simulation with $k - \omega$ turbulent model the velocity of the single jet was still 1.4 m/s (Figure 14.30).

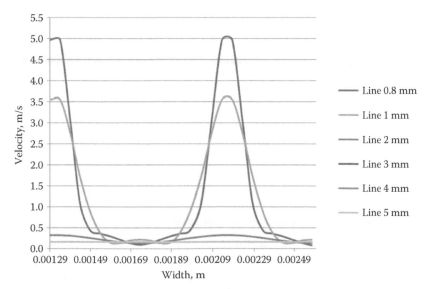

FIGURE 14.29
Velocity profiles, sample 7, case 10: $k - \varepsilon$ turbulent model.

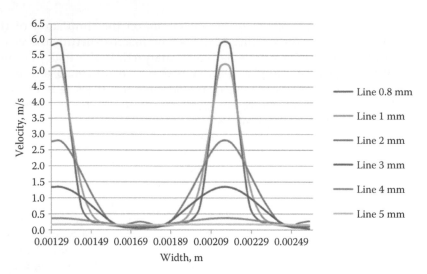

FIGURE 14.30
Velocity profiles, sample 7, case 11: $k - \omega$ turbulent model.

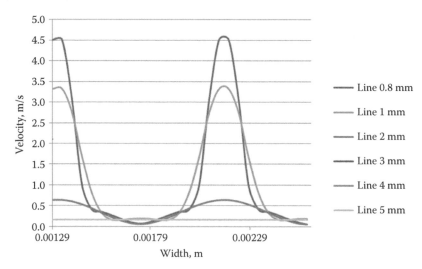

FIGURE 14.31
Velocity profiles, sample 7, case 12: RSM turbulent model.

14.8.2 Effect of the Woven Macrostructure

The assessment of the effect of the characteristics of the meso- and macro-structures of the woven samples on the permeability and flow development was done once again on the basis of the analysis of the velocity profiles for each case. Figures 14.32 and 14.33 visualize the flow development after samples 4 and 10, respectively, in cross sections distanced at 0.4 (0.8), 1, 2,

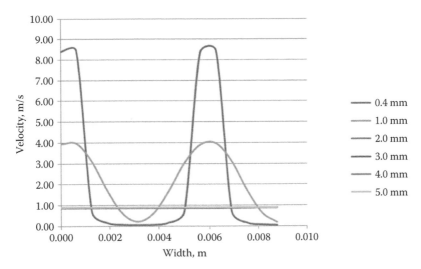

FIGURE 14.32
Velocity profiles, sample 4, case 6: RSM turbulent model.

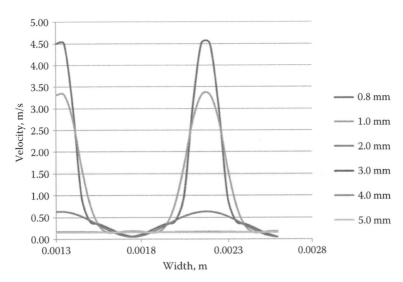

FIGURE 14.33
Velocity profiles, sample 10, case 15: RSM turbulent model.

3, 4, 5 mm from the textile *barrier*. The exemplary cases (case 6 and 15) were simulated by applying RSM turbulent model.

It was clear from the comparison of the velocity profiles that the jet flows and the resulting flow had higher velocities in the case of woven macro-structures with bigger pore area (Figure 14.32). At the same time, the velocity decay was faster for tightly woven samples with lower porosity (Figure 14.33).

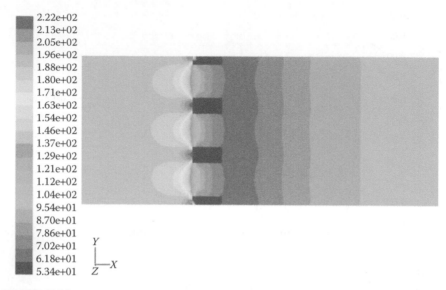

FIGURE 14.34
Static pressure contours, Pa, sample 4, case 4: $k - \varepsilon$ turbulent model.

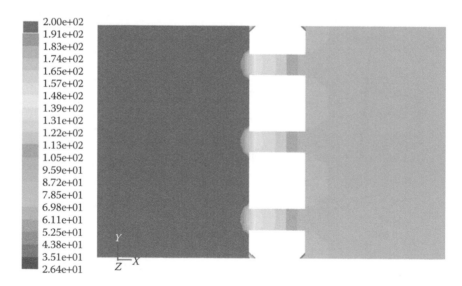

FIGURE 14.35
Static pressure contours, Pa, sample 6, case 7: $k - \varepsilon$ turbulent model.

Figures 14.34 and 14.35 show the numerical results for the static pressure in the computational domain for samples 4 and 6, respectively. The results were obtained by using $k - \varepsilon$ turbulent model. The static pressure changed in the zone of the pores of the macrostructure as expected: it was higher for the sample with lower porosity (sample 6).

14.8.3 Verification of the Numerical Results for the Air Permeability from the Model Study

The verification was done on the basis the numerical results for the mean flow velocity after the woven macrostructure, which corresponded to the air permeability coefficient B_p, determined experimentally. The calculation was performed for each of the 15 cases, simulated with the different turbulent models. The numerical results were compared with experimental data for the air permeability coefficient, presented in Chapter 7. The summary of the comparison is shown in Figure 14.36

The relative error was also calculated for further assessment of the differences between the numerical and experimental results for the air permeability in through-thickness direction of the samples. The results are presented in Table 14.8.

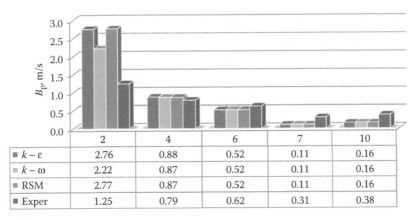

	2	4	6	7	10
$k - \varepsilon$	2.76	0.88	0.52	0.11	0.16
$k - \omega$	2.22	0.87	0.52	0.11	0.16
RSM	2.77	0.87	0.52	0.11	0.16
Exper	1.25	0.79	0.62	0.31	0.38

FIGURE 14.36
Comparison between numerical and experimental results for the air permeability coefficient B_p.

TABLE 14.8

Relative Error between Numerical and Experimental Results for the Air Permeability Coefficient

	Relative Error, %		
Sample	$k - \varepsilon$	$k - \omega$	RSM
2	−120.8	−77.6	−121.6
4	−11.4	−10.1	−10.1
6	16.1	16.1	16.1
7	64.5	64.5	64.5
10	57.9	57.9	57.9

The numerical results for the air permeability coefficient B_p of sample 2 differed in the utmost degree from the experimentally measured values. Actually, this was the most porous macrostructure. The lowest relative error between numerical and experimental results was calculated for the case, simulated with $k - \omega$ turbulent model, but even though the result was unsatisfactory. In fact, the biggest differences between the three turbulent models were registered within the simulation of the air permeability of sample 2. The numerical results for the air permeability coefficient simulated by applying the different turbulent models were equal or very similar for the rest of the samples.

It has to be mentioned that in the work of Saldaeva (2010), where a precise geometric model of an *unit cell* was used to simulate the woven macrostructure (generated by using TexGen specialized software code), the error in simulating the air permeability in through-thickness direction was over 100% for 6 of the 7 samples.

The further analysis of the results in Figure 14.36 showed that the experimentally determined air permeability coefficient for samples 2 and 4 was lower than the simulated one. For the rest of the samples, the experimental values were higher than the predicted values. For samples 4 and 6, however, the relative error was quite low for all turbulent models used: 10% to 12% for sample 4 and 16.1% for sample 6. The discrepancies between the experimental and numerical values for B_p in the case of samples 7 and 10, however, were quite high. Since the samples had a low values of porosity, it was concluded that part of the flow had passed through the pores of the mesostructure (the spaces between the fibers in the yarns), which the simulation model did not account for.

In her study, Saldaeva (2010) commented that the air flow would pass through the larger area of the pores of the macrostructure, and not through the pores of mesostructure. The author was based on this statement when developing the analytical model of the through-thickness air permeability to ignore the transfer of air through the pores between the fibers. In the performed CFD simulation of the air permeability, only the flow through the pores of the macrostructure was simulated.

Though at a first glance this approach seems to be simplified, actually the fluid follows a path that corresponds to the least flow resistance. The experimental measurement of the air transport through the pores of the mesostructure (between the fibers in the threads) only is currently impossible. Behera and Hari (2010) also commented that there was no way to determine the percentage of the air flow through the pores of the threads, but they estimated that it would not exceed more than 10% of the total flow rate in a transverse direction. Moreover, the authors noted that such a high rate would only be valid for very tightly woven macrostructures, where the pore area between the fibers in the yarns was closer to the pore area between the yarns in the fabric. In other cases, the flow rate through the mesostructure would be very low and could be neglected (Behera and Hari 2010).

14.8.4 Modeling of the Porosity of the Mesostructure

It has been already commented in Section I that the computer simulation by means of CFD is one of the few, if not the only opportunity to assess how much of the air that flows in through-thickness direction of a woven macrostructure passes through the pores of the macrostructure and how much—through the pores between the fibers in the mesostructure.

FLUENT CFD software package allows the simulation of the movement of fluid through a porous medium by calculating an additional term in the transport equations. The CFD solver considers the porous medium as a medium with a certain resistance and the additional term is composed of two parts: one that calculates the viscous losses and second that computes the inertial losses. The porosity is set in the range of 0%–100%, and 0% corresponds to a completely impermeable medium.

New cases were simulated for samples 7 and 10, with porosity of the mesostructure from 0% to 5%. Figures 14.37 and 14.38 visualize the flow field of sample 7 with 0% and 5% porosity, respectively. It is obvious that the model reacted to the changed conditions: the development of the flow after the sample was more intense, with increased time of mixing of the jet flows and a much higher velocity of the fluid in the case of porosity of the yarns (Figure 14.38), compared with the case of absence of porosity (Figure 14.37).

Table 14.9 presents the results of the prediction of the air permeability coefficient B_p based on the numerical results for the average flow velocity after

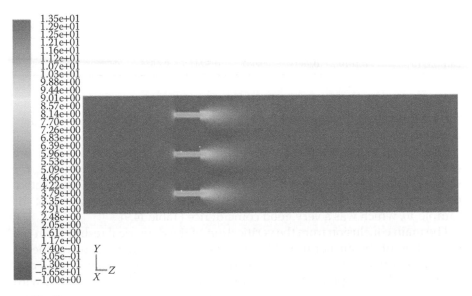

FIGURE 14.37
Velocity contours, m/s, sample 7, 0% porosity, plain $x = 1.29$ mm.

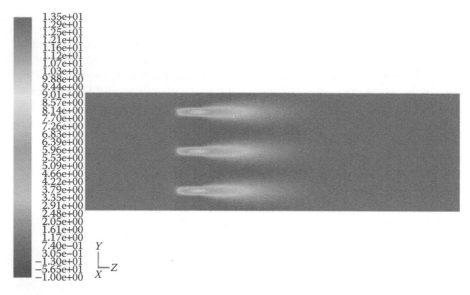

FIGURE 14.38
Velocity contours, m/s, sample 7, 5% porosity, plain $x = 1.29$ mm.

TABLE 14.9

Comparison between Numerical and Experimental Results for the Air Permeability Coefficient B_p

Porosity of the Mesostructure	Air Permeability Coefficient B_p, m/s						
	0%	1%	2%	3%	4%	5%	Experiment
Sample 7	0.1086	0.2345	0.3604	0.4863	0.6122	0.7381	0.31
Sample 10	0.1619	0.2693	0.3967	0.5241	0.6515	0.7789	0.38

the woven macrostructure and their comparison with the experimental data from Chapter 7.

The closest value of the air permeability coefficient was obtained for the cases with 2% porosity of the mesostructure. The predicted values of B_p for samples 7 and 10 were higher than the experimentally obtained. The relative error, compared to the experimental results, was 16% for sample 7 and 4% for sample 10, which was a very good coincidence (Table 14.9).

The main conclusion from the verification of the numerical results from the simulation of the air permeability of woven macrostructures, using the jet systems approach, was that the theoretical results for B_p were in a good correlation with the experiments. Only the numerical result for sample 2, with fabric areal porosity over 50%, differed significantly from the experimental one. For the samples with areal porosity of 20%–30% (samples 4 and 6), the coincidence was very good without simulation of air transfer through the mesostructure. For samples with areal porosity in the range 0%–15% (samples 7 and 10), the match was very good after simulation of air flow through the fibers in the mesostructure. It was found that for samples 7 and 10, 98% of the measured flow rate in Chapter 7 has passed through the pores between the threads and 2% of the flow rate has passes through the pores of the fibers in the yarns.

14.9 Summary

The necessity of computer simulation of the air transfer through woven macrostructure by means of CFD was discussed. The methods for computer simulation were analyzed.

An original approach for simulating fluid flow through a woven macrostructure based on the theory of jet systems was presented. Such an approach was not applied to date in the literature.

The numerical procedure for the parametric study used to determine the conditions of the numerical simulation of the transmission of fluids and heat in woven macrostructure were presented. The conditions and numerical procedure for the model study of the air permeability of real woven macrostructures, subject of research in Chapter 7, were also detailed.

The numerical results from the systematic parametric study were presented. It was found that a system of in-corridor-ordered 3 × 3 jet system was representative for a system of countless jets of the same type. Thus, the already known theory for jet systems was proven numerically.

It was shown that the simulation of the air permeability of woven macrostructures through a system of in-corridor-ordered 3 × 3 jet system led to reliable quantitative and qualitative results; thus, there was no necessity to create a virtual model of the macrostructure with a larger number of threads/pores.

It was found that the approximation of the shape of the cross section of the pores in the woven macrostructure to a circle or a square is not important for the theoretical calculations and numerical studies, as far as the pore area in the model corresponded to the real pore surface. Such a result was not reported so far by other studies.

It was found that structured grids—with and without local refinement—described better the flow decay. It was also found that the use of a standard $k - \varepsilon$ and RSM turbulent models yielded to obtaining of comparable results in terms of the development of the flow and its velocity.

The numerical results from systematic model analysis of the air permeability of real woven macrostructures, subject of the investigation in Chapter 7, were presented. The effect of characteristics of the meso- and macrostructure of the woven samples on the fluid flow in through-thickness direction was proven.

It was proven that the simulation of the transmission of air through a woven macrostructure using the jet systems approach gave similar results to the experiment with the exception of very porous macrostructures. For samples with areal porosity of 20%–30%, the coincidence between the numerical and experimental results for the air permeability was very good without simulating the transfer of air through the mesostructure. For samples with areal porosity in the range of 0%–15%, the coincidence was very good after simulating the transmission of air through the mesostructure.

15

Mathematical Modeling and Numerical Simulation of Heat Transfer through Woven Structures by CFD

15.1 Mathematical Model

The Reynolds-averaged Navier–Stokes partial differential Equations 15.1 through 15.3 were used for the simulation of the heat transfer through a thin woven macrostructure. The energy Equation 15.5 together with the continuity Equation 15.4 was added to the system.

$$\frac{\partial u}{\partial t} + u\frac{\partial u}{\partial x} + v\frac{\partial u}{\partial y} + w\frac{\partial u}{\partial z} = X - \frac{1}{\rho}\frac{\partial p}{\partial x} + \nu\left(\frac{\partial^2 u}{\partial x^2} + \frac{\partial^2 u}{\partial y^2} + \frac{\partial^2 u}{\partial z^2}\right) -$$

$$\rho\left(\frac{\partial \overline{u'}^2}{\partial x} + \frac{\partial \overline{u'v'}}{\partial y} + \frac{\partial \overline{u'w'}}{\partial z}\right) \tag{15.1}$$

$$\frac{\partial v}{\partial t} + u\frac{\partial v}{\partial x} + v\frac{\partial v}{\partial y} + w\frac{\partial v}{\partial z} = Y - \frac{1}{\rho}\frac{\partial p}{\partial y} + \nu\left(\frac{\partial^2 v}{\partial x^2} + \frac{\partial^2 v}{\partial y^2} + \frac{\partial^2 v}{\partial z^2}\right) -$$

$$\rho\left(\frac{\partial \overline{u'v'}}{\partial x} + \frac{\partial \overline{v'}^2}{\partial y} + \frac{\partial \overline{v'w'}}{\partial z}\right) \tag{15.2}$$

$$\frac{\partial w}{\partial t} + u\frac{\partial w}{\partial x} + v\frac{\partial w}{\partial y} + w\frac{\partial w}{\partial z} = Z - \frac{1}{\rho}\frac{\partial p}{\partial z} + \nu\left(\frac{\partial^2 w}{\partial x^2} + \frac{\partial^2 w}{\partial y^2} + \frac{\partial^2 w}{\partial z^2}\right) -$$

$$\rho\left(\frac{\partial \overline{u'w'}}{\partial x} + \frac{\partial \overline{v'w'}}{\partial y} + \frac{\partial \overline{w'}^2}{\partial z}\right) \tag{15.3}$$

$$\frac{\partial u}{\partial x} + \frac{\partial v}{\partial y} + \frac{\partial w}{\partial z} = 0 \qquad (15.4)$$

$$\rho\frac{De}{Dt} = \rho\dot{q} + \frac{\partial}{\partial x}\left(\lambda\frac{\partial T}{\partial x}\right) + \frac{\partial}{\partial y}\left(\lambda\frac{\partial T}{\partial y}\right) + \frac{\partial}{\partial z}\left(\lambda\frac{\partial T}{\partial z}\right) - p\left(\frac{\partial u}{\partial x} + \frac{\partial v}{\partial y} + \frac{\partial w}{\partial z}\right) +$$

$$\mu_v\left(\frac{\partial u}{\partial x} + \frac{\partial v}{\partial y} + \frac{\partial w}{\partial z}\right)^2 + \mu\left[2\left(\frac{\partial u}{\partial x}\right)^2 + 2\left(\frac{\partial v}{\partial y}\right)^2 + 2\left(\frac{\partial w}{\partial z}\right)^2 + \left(\frac{\partial u}{\partial y} + \frac{\partial v}{\partial x}\right)^2\right] + \quad (15.5)$$

$$\left(\frac{\partial u}{\partial z} + \frac{\partial w}{\partial x}\right)^2 + \left(\frac{\partial v}{\partial z} + \frac{\partial w}{\partial y}\right)^2\right]$$

where:
   e is the internal energy, W
   μ_v is the volume viscosity coefficient

Thus, the physical phenomenon was described by three fundamental physical laws: conservation of mass law, conservation of energy law (first law of thermodynamics), and Newton's second law.

15.2 Numerical Procedure for the Study

The numerical procedure for the modeling of heat transfer in through-thickness direction of a woven macrostructure coincides to a great extent with the basic formulations of the model study of the air permeability, presented in Chapter 14. The simulation was performed with the commercial CFD software package FLUENT, and the geometry of the woven macrostructure was built with the FLUENT preprocessor GAMBIT. The control volumes method was used as a method for discretization. SIMPLE algorithm was again applied to solve the system of algebraic equations.

The jest system approach was used for the simulation of the heat transfer. The woven macrostructures were approximated to a 3 × 3 in-corridor ordered jet system.

15.2.1 Computational Domain

The woven macrostructure was built by elements with a size of $0.225 \times 1.365 \times 1.365 \ mm^3$. The textile was placed at the inlet of the computational domain. The length of the computational domain after the woven macrostructure was 5 mm.

TABLE 15.1

Characteristics of the Simulated Woven Macrostructures

Sample	P_{wa}, threads/dm	P_{wf}, threads/dm	Tt_{wa}, tex	Tt_{wf}, tex	Thickness, mm	Weave	Fabric Weight, g/m²	Fabric Cover Factor, %	Equivalent Side of the Pore, a_{eqv}, mm
6	234	254	30	30	0.47	Plain	157	76.69	0.2136
7	383	338	28	28	0.68	Twill 2/1Z	184	86.95	0.2500
10	386	286	30	36	0.71	Twill 3/1Z	202	91.12	0.1558

15.2.2 Selection of the Woven Macrostructure

Three woven macrostructures for clothing and bedding were chosen for the simulation: samples 6, 7, and 10, which were studied experimentally in Chapter 7. Their air permeability was simulated in Chapter 14. The main characteristics of the macrostructures are shown in Table 15.1. All samples were woven of cotton yarns, which eliminated the effect of the type of the material, set as an initial condition.

15.2.3 Jet System and Turbulence Modeling

A 3 × 3 in-corridor ordered jet system was applied. The RANS system was closed by using $k - \varepsilon$ turbulent model.

15.2.4 Selection of a Grid

A structured grid was generated. Figure 15.1 illustrates a cross section of the computational grid for sample 6. Figure 15.2 is a view of the computational domain for the simulation of the heat transfer through sample 7. Data for the generated grids are summarized in Table 15.2.

15.2.5 Initial and Boundary Conditions

The type of the material was set with the corresponding data for the simulation: density of the material, its specific heat, and thermal conductivity. For the simulation cases with convective flow, the speed of the convective flow was set to 0.5 m/s. Symmetric boundary conditions were set on the walls of the computational domain. The pressure gradient was 100 Pa.

Temperature gradient of 14°C was set between the two sides of the woven macrostructure: 36°C from the *hot* side and 22°C from the *cold* side. Thus, the real physical situation of heat transfer between the skin of a human body and

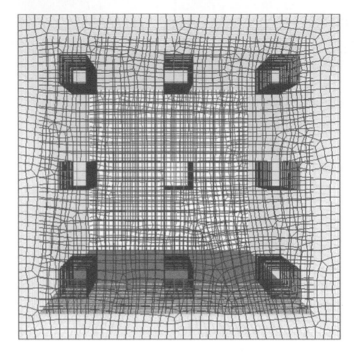

FIGURE 15.1
Cross section of the computational grid, sample 6.

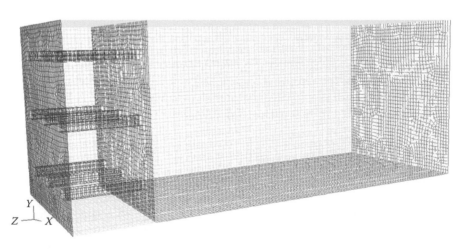

FIGURE 15.2
Computational domain, sample 7.

TABLE 15.2

Grid Data

Sample	3D Cells	Faces	Nodes
6	214,533	660,570	231,620
7	310,400	952,240	331,539
10	608,634	1,858,923	641,772

the indoor environment with a standard temperature (in accordance with ISO 7730, 1995) was simulated.

15.2.6 Convergence Conditions

The absolute value of the residuals was set to 1.10^{-6} for the continuity equation and 1.10^{-5} for the rest of the values.

15.2.7 Summary of the Investigated Cases

For each of the woven macrostructures, the following cases were studied:

- Solid structure
- Porous structure
- Cases without a convective flow
- Cases with a convective flow in transversal direction of the heat transfer

Table 15.3 summarizes the cases of the numerical simulation of the heat transfer. Nine cases were simulated all together.

TABLE 15.3

Simulated Cases—Heat Transfer

Case	Sample	Solid Structure	Jet System	Convective Flow
1	6	Yes	–	–
2	6	–	3×3	–
3	6	–	3×3	Yes
4	7	Yes	–	–
5	7	–	3×3	–
6	7	–	3×3	Yes
7	10	Yes	–	–
8	10	–	3×3	–
9	10	–	3×3	Yes

15.3 Numerical Results

Figure 15.3 shows an example of the results for the contours of the static temperature for case 4 (sample 7). The macrostructure was simulated as a *solid* one, that is, the presence of air gaps between the threads was not taken into account. The respective distribution of the static temperature, but for the approximation of the macrostructure as an in-corridor ordered jet system, is shown in Figure 15.4 (case 5, sample 7) and Figure 15.5 (case 8, sample 10). The visualization of the heat transfer in through-thickness direction of the woven macrostructure showed logical development of the physical process for all cases.

The simulation of the heat transfer through the woven macrostructure in the presence of a convective flow in the direction, transverse to the heat transfer through the macrostructure, was of a particular interest. Actually a natural convection flow is available around the human body with the direction of motion from the floor to the ceiling.

Figures 15.6 through 15.8 visualize the heat transfer through the investigated samples in the presence of a convective flow near the sample: cases 3, 6, and 9, respectively. The visible difference in the temperature distribution was most likely due to the effect of the macrostructure of the studied

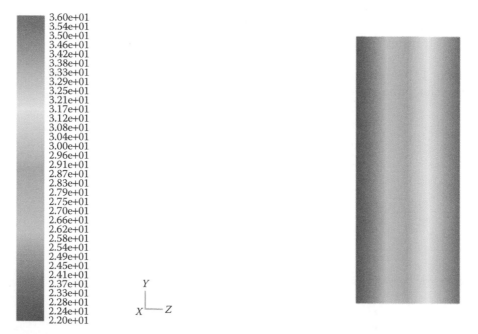

FIGURE 15.3
Contours of static temperature, °C: sample 7, case 4.

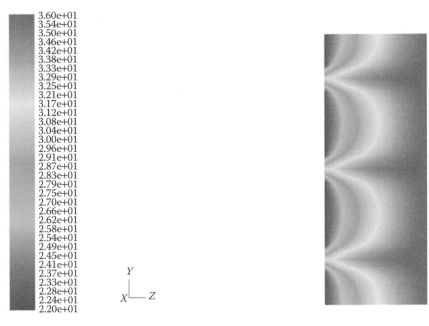

FIGURE 15.4
Contours of static temperature, °C: sample 7, case 5.

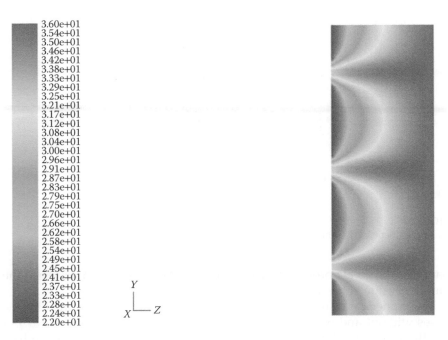

FIGURE 15.5
Contours of static temperature, °C: sample 10, case 8.

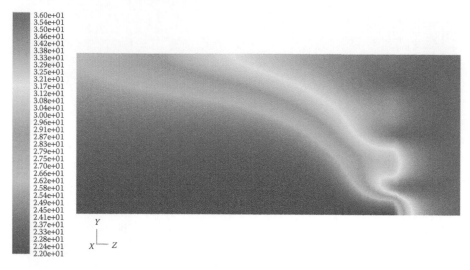

FIGURE 15.6
Temperature field, °C: sample 6, case 3.

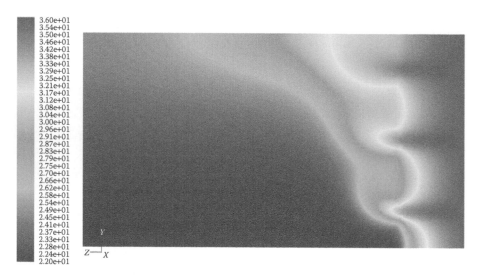

FIGURE 15.7
Temperature field, °C: sample 7, case 6.

samples: the areal porosity of sample 6 was highest, while sample 10 had the smallest value of the areal porosity.

The numerical results for the heat transfer and heat flux, obtained for the cases without and with simulation of porosity, are compared in Table 15.4 for the three investigated woven macrostructures. The values for the heat transferred through a solid macrostructure were higher in comparison with

FIGURE 15.8
Temperature field, °C: sample 10, case 9.

TABLE 15.4

Heat Transfer—Numerical Results

Sample	6		7		10	
Area, m²	0.0000037249		0.0000057600		0.0000066564	
Case	Heat Transfer, W	Heat Flux, W/ m²	Heat Transfer, W	Heat Flux, W/ m²	Heat Transfer, W	Heat Flux, W/m²
Solid	3.472E–05	9.3209	4.562E–05	7.9194	5.151E–05	7.7390
Porous	1.238E–05	3.3236	7.536E–06	1.3083	6.4E–06	0.9615

the respective values for the porous structure. This result is explained by the thermal insulation properties of cotton as compared to air: the thermal conductivity coefficient for cotton is 0.04 W/(mK), while for the air is 0.024 W/(mK) (Gebhart 1993). The presence of air voids between the threads has a function of a better insulator.

This result confirms the known fact that the main function of the textiles is to keep an air layer close to the body, since it is the air that acts as a thermal insulator. The visualization of the interaction with a transverse convective flow (Figures 15.6 through 15.8) showed how the air between the pores of the macrostructure can enter into interaction with a flow around the human body. No matter if this flow will be a result of natural or forced convection, the air layer between the pores of the macrostructure can be *removed* at the expense of reduced thermal insulation of the clothing.

15.4 Summary

Results from simulation study of the heat transfer in through-thickness direction of a woven macrostructure were presented.

Numerical results from different cases were compared: for heat transfer of a woven macrostructure, modeled as a solid structure, as a system of in-corridors ordered 3 × 3 jet system, and in the presence of a convective flow in the transverse direction of the heat transfer through textile barrier.

The simulation confirmed the effect of the presence of a stationary air layer in the near-textile region that increased the thermal insulation properties of the textile layers.

Section IV

Mathematical Modeling and Numerical Study of Thermophysiological Comfort with a Thermophysiological Model of the Human Body

16

Thermoregulation of a Clothed Body: Physiological Peculiarities

As a part of the clothing, the textile macrostructures provide specific, mobile and dynamically changing microenvironment around the human body. Thus, they support the thermoregulatory processes in the human body and the achievement of thermophysiological comfort.

In Section IV of the book, results from the modeling and numerical study on the thermophysiological comfort of the human body were presented. The following topics were considered:

- Physiological peculiarities of the thermoregulation of the human body and the possibilities for modeling of the thermoregulation by thermophysiological model (Chapter 16).
- Numerical study of the thermophysiological comfort of a clothed body (Chapter 17).
- Results from the numerical study of the thermophysiological comfort of a clothed body (Chapter 18).

The main aim of this chapter is to systematize the fundamental prerequisites for modeling and numerical study of the thermophysiological comfort of clothed body in different thermal environment. The role of the hypothalamus and mechanisms for thermoregulation of the body in hot and cold environment are analyzed from the viewpoint of ensuring thermophysiological comfort.

The development of thermophysiological models is discussed. The advantages and details of the Gagge's thermophysiological model are presented.

16.1 Thermophysiological Control of the Human Body

The human body has a complex mechanism for temperature control, which has to balance both heat production and heat losses. The heat is generated continuously as a result of the metabolism of all body cells, but it changes with the activity of the glands and the internal organs, as well as the action of the muscles, which movement (i.e., shivering) is one of the effective ways of the body to increase the heat generation.

16.1.1 Role of the Hypothalamus

The hypothalamus in the brain is the gland that controls the body temperature. It receives two types of signals (Parsons 2003):

- From the peripheral nerves, which register the sensation of receptors for cold/warmth and send a signal to the front part of the hypothalamus. The receptors for cold/warmth are particularly sensitive: those for warmth are able to register temperature increase of 0.007°C and the cold receptors can register temperature decrease of 0.012°C.
- From the temperature of the blood, which irrigates the area of the hypothalamus and provides signals to the back part of the hypothalamus.

The two types of signals are integrated into the center of the hypothalamus so as to assure the maintenance of normal temperature of the body and particularly—of the core body. The way of integration of the two signals is still unknown. The main task of thermoregulation, however, is to maintain core body temperature constant at around 37°C, while the presence of a sensation for warmth or cold on the skin surface is considered and treated as *discomfort* only (Havenith 2002).

Figure 16.1 summarizes the role of the hypothalamus for controlling the temperature inside the body.

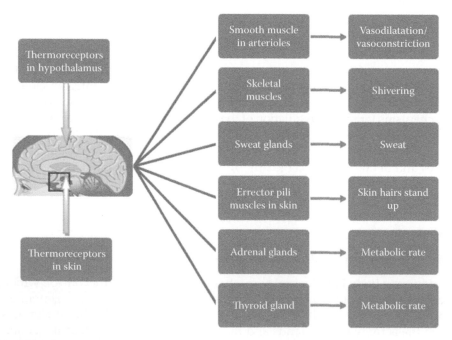

FIGURE 16.1
Role of the hypothalamus for control of the body temperature.

If the thermoreceptors in the region of the hypothalamus detect an increased blood temperature or core body temperature, respectively, the output signals from the hypothalamus provoke vasodilatation of the blood vessels in the skin and sweat appearance, thus increasing the heat losses. If the body temperature decreases, the heat losses are limited by the vasoconstriction of peripheral blood vessels, and the production of heat increases simultaneously. Both vasoconstriction and vasodilatation are caused by contraction of the smooth muscle tissue, which is located in the middle layer of the walls of the arterioles. At the same time, the heat production is augmented by shivering of the muscles (uncontrolled muscles' contraction), increased metabolism via reaction of the adrenal or thyroid glands, and so on. The erector pili muscles, located below the skin surface, cause pricking so as the skin hair to retain a larger volume of motionless air close to the skin (Huizenga et al. 2001; Havenith 2002; Parsons 2003; Fiala et al. 2010).

16.1.2 Mechanisms of Body's Thermoregulation

The thermoregulation of the human body has to ensure proper functioning of all organs and systems independently of the temperature of the environment. Different thermoregulatory mechanisms are set in motion depending on the environment around the body: hot or cold (Figure 16.2).

16.1.2.1 Effect of the Environment with High Air Temperature

Provided that the temperature of the skin is higher than the temperature of the environment, it is necessary the temperature of the body to be decreased by the process of heat losses, which uses two mechanisms: vasodilatation

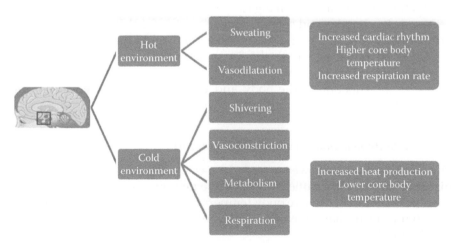

FIGURE 16.2
Thermoregulatory mechanisms depending on the type of environment.

and sweating. Vasodilatation leads to expansion of blood vessels in the periphery of the body, through which the blood moves away from the internal organs to the periphery (skin). At the same time, the heart rate increases to accelerate the transfer of blood to the extremities and skin surface. Thus, heat losses through the skin by radiation, conduction, and convection appear. Very effective are perspiration and evaporation of sweat from the skin that detract heat by accelerating the heat losses. The last again leads to decrease of the core body temperature and avoiding of hyperthermia.

16.1.2.2 Effect of the Environment with Low Air Temperature

When exposed to low temperatures, the body maintains its core temperature via vasoconstriction, increased heat production, and behavioral changes.

Avoiding hypothermia requires the activation of mechanisms for heat retention in the core body, that is, increase in the heat production by metabolic processes in the muscles and liver, while reducing heat losses from the skin and lungs. The vasoconstriction induces constriction of blood vessels in the periphery of the body, which provokes a substantial portion of the blood to circulate in the torso and internal organs, thus reducing the blood flow to the extremities. This process can increase the body temperature of 1°C–2°C due to the reduction in heat losses through the three mechanisms of heat transfer: conduction, convection, and radiation. Vasoconstriction creates a feeling of cold, causing muscle tremors and increased heat production. Heat production from the liver also adds to the process. The behavioral reaction of the individual to add additional layers of clothing or bedding also contributes to the augment of the core body temperature and the reduction in heat losses from the skin. Thus, when exposed to cold environment, the body maintains its internal temperature via vasoconstriction, increased heat production, and behavioral changes.

Despite its complex and sensitive mechanisms for thermoregulation, the human body is coping with the aggressive impact of the environment in a relatively narrow range. Both the increase in the core body temperature, hyperthermia, and the reverse process, hypothermia, have an adverse effect on the function of the body and can lead to severe disability and even death.

16.1.2.3 Body Reactions of Thermal Discomfort

Out of these threats, however, even in cases, when the protective mechanisms of the body are sufficiently effective and thermoregulatory processes are able to maintain the core body temperature at about 37°C, the cold or hot environment can cause a sequence of events associated with *thermal discomfort* (Figure 16.3).

Both high and low temperature, that is, of the working environment, cause delay in reaction time. While the hot ambient causes fatigue, however, the

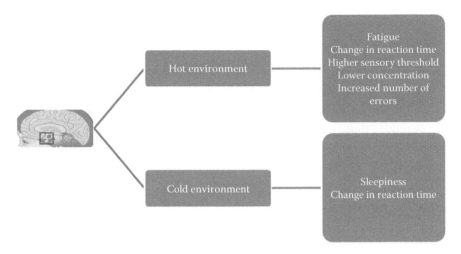

FIGURE 16.3
Physiological reactions of thermal discomfort.

low temperature of the environment increases sleepiness of the individual. The two effects are, to their nature, pure physiological processes, that is, the accelerated heart rate causes fatigue and the lack of irrigation, including in the brain tissue, causes sleepiness. In a hot environment, the higher threshold of sensory perception, coupled with decreased concentration, is also observed. All these result in a larger number of subjective errors.

16.2 Thermophysiological Models

Thermophysiological models try to describe the thermoregulatory mechanisms of the human body as closer to their biological nature as possible. The work on this topic started during the first half of the twentieth century, but the first models appeared after the 1950s.

Thermophysiological models from the 1960s were based on their predecessors: the temperature scales for determining the thermal comfort of the environment. The first scale of that type was described in the work of Houghton and Yaglou (1923). Its development aimed the scale to be included in the first edition of Heating and Ventilating Guide established in 1922 American Society of Heating and Ventilating Engineers, today ASHRAE. Developed on the basis of precise experimental studies, the first temperature scale for thermal comfort was continuously used for decades. It showed the importance of temperature and humidity of the environmental air for people's sense of comfort and perceived standard of working conditions for many professions (Lind 1970).

Every thermophysiological model should include two types of processes:

- The thermal processes, which include description of two thermal interrelations:
  - Thermal exchange inside the body.
  - Thermal exchange between the body and environment (both indoor and outdoor).
- The mechanisms, used by the body to maintain its temperature (core body temperature) within a narrow range of values.

16.2.1 Hardy's Classification

There are several classifications of thermophysiological models in the literature (Hardy 1972; Houdas 1981; Lotens 1988; D'Ambrosio 2006). One of the most frequently used is the Hardy's approach, where the models are divided into two groups in accordance with their nature: qualitative and quantitative (Hardy 1972).

16.2.1.1 Qualitative Models

These are descriptive models, with little application, but have contributed to the emergence of the quantitative models. It is worth the following quality models to be mentioned:

- *Verbal models*: They have been created as a first attempt to describe the human thermoregulatory system. They were based on different hypotheses and were too descriptive and simple to have success.
- *Figurative models*: They were a step further toward the description of the complicated thermoregulatory system of the human body. The figure-based descriptions were more detailed than the verbal models. However, they still consider the thermoregulatory system as a black box, without ability for describing and analyzing the complex interrelations between the single model components.
- *Neuronal models*: They considered the human thermoregulation only at nervous system level, without dealing with other systems and organs like vascular systems, muscles, and glands. Being limited in their *input–output* reactions, they had constricted application.
- *Chemical models*: They appeared on the basis of the developments in the nervous system microanatomy. The chemical models considered the exchange of information as production and exchange of chemical substances in the human body. Despite the innovative approach, the chemical models were again limited.

16.2.1.2 Quantitative Models

By contrast with the qualitative models, the quantitative models allow prediction and/or calculation of different quantities on the basis of already established quantitative interrelations. The basic models of the group are as follows:

- *Physical models*: They are based on the idea that the human body is a thermodynamic system, which interacts with the environment. However, the human body is again treated as a black box. Being poorly versatile and quite expensive, they are not used nowadays.
- *Mathematical models*: They built the mathematical background of the modern simulation models and software, finding mathematical description of human–environmental interaction expressed in heat exchange between them. Among the most popular mathematical models are these of Winslow et al. (1936, 1937), Gagge (1937), and Gagge et al. (1938). The main disadvantage of the mathematical models is the high level of abstraction, widely compensated by their versatility. The first principle of thermodynamics actually has formed the background of the mathematical description as follows:

$$S_b = M \pm W \pm R_s \pm C \pm K \pm Q_e \qquad (16.1)$$

where:
S_b is the rate of the body heat storage, W
M is the metabolic rate, W
W is the effective mechanical power, W
R_s is the radiative heat flow, W
C is the convective heat flow, W
K is the conductive heat flow, W
Q_e is the evaporative heat flow at the skin surface, W

16.2.2 Classification Based on the Number of Body Segments

Another option for classifying the thermophysiological models is the number of parts of the human body included in the model. Such models occurred as a result of the development of mathematical modeling and the growing importance of the computer simulation of the phenomenon of *thermal regulation of the human body*.

16.2.2.1 One-Cylinder Models

They are defined as one-cylinder models as the human body is presented as a homogeneous cylindrical body, which is approximated to a thermodynamic system. The first model of that type was developed by Burton (1941), in which Bessel functions were used to describe the thermodynamic phenomena. Later on, a series of one-cylinder models were proposed and validated by means of experiments (D'Ambrosio 2006).

16.2.2.2 Two-Cylinder Models

One of the most important among the models of the 1970s is the Gagge's model (Gagge et al. 1971; Gagge 1973), which was developed further by the author and several other researchers (Gagge et al. 1986, 1996; Fobelets and Gagge 1988).

The most important feature of Gagge's model is the description of the human body as two-layered cylinder, having as an inner cylinder the core body and as an outer cylinder the skin surface (epidermis). Each layer has its own energy balance, and at the same time, it is in a thermal exchange with the other layer and the surroundings.

The core body uses conduction and convection to exchange thermal energy with the skin layer, taking into account the real function of the cardiovascular system (blood vessels), the generated energy by the muscles, and the metabolism. At the same time, the core body exchanges heat with the surrounding environment through respiration. The outer layer uses evaporative and dry heat transfer (through convection and radiation) to exchange heat with the environment.

Though there are more complicated models, the Gagge's model was used in the development of the theory for calculating the ET* index (new effective temperature index) on which the ASHRAE 55-92 standard was based on (Fobelets and Gagge 1988; ASHRAE 1992).

It is especially suitable for simulation of heat transfer through a layer of clothing, because the model allows an extra layer to be placed on the outer layer of the skin's epidermis.

16.2.2.3 Multilayered and Multisegment Models

After the Gagge's model development, it was realized that the one-cylinder type of models was not enough to simulate the thermoregulatory system behavior and, at the beginning of 1970s, new models, based on more cylinders, were created.

Wissler (1961, 1963) proposed a new multilayered model, suggesting all physical and physiological factors of the model to be taken into account, that is, local differences in temperature, convective heat transfer through blood vessels, heat losses by evaporation through skin and respiratory system, and thermal storage inside the body.

One of the most important models among the developments in this group is the model of Stolwijk (1970). This is the first model able to take into account all factors, mentioned by Wissler (1961), and therefore, it was suggested for NASA Skylab and Apollo programs.

Stolwijk's model presents the human body as a six-part model, formed by head, trunk, arms, hands, legs, and feet. Each part has four layers: core, muscles, fat, and skin. A control system for commands and exchange of information is used, together with a controlled system for every single layer of the body—25 control systems all together. The main advantage of Stolwijk's model is its completeness. It has summarized the knowledge on thermoregulation in human body in the time of its development.

16.2.2.4 New Models

Despite the variety of modern thermophysiological models, they use as a background either the Gagge's model or the Stolwijk's model. However, the new models go further in using more sophisticated methods to predict the human body's thermoregulation. Havenith's model (Havenith 2001) is an evolution of Gagge's model, which takes into account physical characteristics of the subject (body surface, fat percentage, etc.). Tanabe's model (Tanabe et al. 2002) uses the approach of Stolwijk's model, but it is already more complicated with 16 body segments and 65 nodes. Similar are the models of Fiala (Fiala et al. 1999) and Schellen (Schellen et al. 2013). An exception is the model of Holopainen (2012), based on the predicted mean vote (PMV) model of Fanger, which is a multisegment model, created for integration with commercial software package for building simulation. The model is verified, but it is not implemented in the software package.

All authors of new models of human thermoregulation reported that the models were validated through experimental results, taken from experiments either with manikins or with human subjects. However, in the work of Jones (2002) it was demonstrated that two different models can predict different results under the same environmental conditions. Therefore, the integration between a thermophysiological model and a software package for dynamic assessment of the environment is a potential solution of the problem.

16.3 Gagge's Thermophysiological Model

16.3.1 Advantages of the Model

The choice of the Gagge's thermophysiological model (Gagge et al. 1971) for the simulations in this part of the study was based on the following advantages of the model:

- The model allows a layer of textile to be added over the skin layer.
- The model underlies the ASHRAE 55-1992 standard "Thermal Environmental Conditions for Human Occupancy," adopted by ANSI (American National Standards Institute).

- The model has been integrated with the FLUENT CFD software package by Pichurov (2009), presented also in Pichurov and Stankov (2013) and Pichurov et al. (2014).

- The model has been widely verified with experimental data and used in various engineering tasks related to thermal comfort and evaluation of the indoor environment in premises.

16.3.2 Description of the Model

Gagge's thermophysiological model describes a standard human (male) body with a body weight of 81.7 kg and a height of 177 cm, corresponding to DuBois body surface $A = 2.00$ m^2, determined in accordance with the formula (DuBois and DuBois 1916; Wang et al. 1992)

$$A = \left(W_b^{0.425} H_b^{0.725} \right) \times 0.007184 \qquad (16.2)$$

where:
  W_b is the weight of the body, kg
  H_b is the body height, cm

The human body is described as two concentric layers (cylinders), one of which forms the core of the body and the other, thinner, is the skin.

The heat exchange between the body and the environment, managed by the thermoregulatory system, is carried out continuously through the skin surface. At the same time, heat is continuously produced inside the body (core body) through various biochemical reactions or muscles work. The heat generated in the core of the body is transmitted by convection to the surface of the skin through the bloodstream and heat exchange in radial direction. The heat from the skin is transferred to the environment by convective heat transfer, radiation, and latent heat transfer (evaporation of sweat or water vapor). The excess heat, which cannot be removed, is *stored* in the core of the body, thereby increasing its temperature (Gagge et al. 1971)

The Gagge's model uses seven independent variables:

- Rate of the body metabolism
- Work accomplished
- Conductive heat transfer coefficient
- Combined heat transfer coefficient for convection and radiation
- Clothing insulation
- Environmental temperature
- Relative humidity

The model includes two systems: passive and control (Gagge et al. 1971; Li and Holcombe 1998).

16.3.3 Passive System

16.3.3.1 Heat Exchange between the Body and the Environment

The equation for the heat balance between the body and the surrounding air stays in the bottom of the Gagge's model (Gagge et al. 1971):

$$S_b = M - Q_e \pm R_s \pm C - W \qquad (16.3)$$

The heat stored in the body, S_b, is defined as warming (if $S_b > 0$) or cooling (if $S_b < 0$). The radiant heat flow R_s represents the heat generated by the body (if $R_s > 0$) or the heat losses by radiation (if $R_s < 0$). The convective heat flux C reflects the heat generated by the body (if $C > 0$) or the convective heat losses (if $C < 0$).

The speed of the metabolic processes M is proportional to the oxygen consumption, which can be measured directly. The total losses of the latent heat transfer can be determined experimentally by weighing and accurate measurement of the lost weight or to be estimated from the following expression:

$$Q_e = E_{res} + E_{dif} + E_{rsw} \qquad (16.4)$$

where:
  E_{res} represents the heat losses by evaporation of moisture from the lungs during exhalation, W/m^2
  E_{dif} represents the heat losses by diffusion of water vapor through the skin surface, W/m^2
  E_{rsw} represents the heat losses by evaporation of sweat during body's thermoregulation, W/m^2

The sum $(E_{res} + E_{dif})$ is determined as *insensible* heat losses, while E_{rsw} represents the *sensible* heat losses from the body.

The losses E_{res} are, in general, proportional to the gradient of the vapor pressure between the lungs and the surrounding air and the ventilation rate of the lungs. The ventilation of the lungs in turn is proportional to the metabolic rate. An equation, derived by Fanger (1967), can be used for calculation of E_{res}:

$$E_{res} = 0.0023M(44 - \phi_a P_a), W/m^2 \qquad (16.5)$$

where:
  44 mmHg is the saturated vapor pressure of the lungs for a mean lungs temperature of 35.5°C
  ϕ_a is the air relative humidity, %
  P_{sa} is the saturated vapor pressure for temperature of the surrounded air T_{sa}, mmHg

The maximum value of the evaporative heat losses E_{max} is calculated from the relation (Gagge et al. 1969):

$$E_{max} = LR.h_c \left(P_{sk} - \phi_a P_a \right) F_{pcl}, \text{W/m}^2 \tag{16.6}$$

where:

LR is the Lewis relation, which takes into account both the diffusion of heat and mass, °C/kPa

h_c is the convective heat transfer coefficient, W/m²

P_{sk} is the saturated vapor pressure for a mean skin temperature T_{sk}, mmHg

F_{pcl} is the coefficient of effective permeability from the skin surface to the environment through a textile layer

The heat losses due to evaporation from the skin surface E_{sk}, W/m², are calculated in accordance with

$$E_{sk} = E_{dif} + E_{rsw} \tag{16.7}$$

The ratio (E_{sk}/E_{max}) has been used by Belding and Hatch (1956) as an index of *environmental heat stress*.

To calculate the ratio (E_{dif}/E_{max}), the Gagge's model uses the relation from the investigation of Brebner et al. (1956):

$$w_{dif} = \frac{E_{dif}}{E_{max}} = 0.06 \quad \text{if} \quad E_{rsw} = 0 \tag{16.8}$$

where:

w_{dif} is the dampness on the skin surface due to diffusion

The values of w_{dif} change from 0.06, when there is no sweating from the skin (considered as normal dampness factor), to 0.04 in case of dehydration of the skin if conditions of cold environment with low humidity are combined.

The heat losses by water vapor diffusion through the skin E_{dif} and those from evaporation of perspiration due to the thermal regulation E_{rsw} do not occur in the same section of the epidermis at the same time (Gagge et al. 1971). At high temperatures of the environment, resulting in the activation of the sweat glands, the separated secretion covers the surface of the skin as a thin film over the whole body, so that the areas with insensible thermal diffusion losses are reduced and those in which evaporation of the sweat begins to cool the body—increase in number and size. So at any point in time it is met:

$$w_{rsw} = \frac{E_{rsw}}{E_{max}} \tag{16.9}$$

where:

w_{rsw} is the dampness of the skin as a result of the sweating

When the body surface is fully wet, $w_{rsw} = 1$. Thus, the two limits of the model with respect to the skin dampness w are set:

- In the absence of thermal regulation by sweating, $w = w_{dif} = 0.06$.
- In the case of thermal regulation by sweating, $w = w_{rsw} = 1$.

These limits, as well as the intermediate values, are described under the condition that the total heat losses from the skin surface E_{sk} are

$$E_{sk} = (0.06 + 0.94w_{rsw})E_{max} \tag{16.10}$$

In the model of Gagge (Gagge et al. 1971), the temperature of the environment T_a is assumed to be constant, and therefore, the following relationship is used for the dry heat exchange:

$$(R_s + C) = h(T_{sk} - T_a)F_{cl} \tag{16.11}$$

where:
h is the combined heat transfer coefficient, W/m²
$h = h_r + h_c$, W/m²

where:
h_r is a sum of the linear radiation exchange coefficient
h_c is the convective heat transfer coefficient
F_{cl} is the dry heat efficiency factor that reflects the effectiveness of the transfer of the dry heat from the skin with surface temperature T_{sk} to the environment with air temperature T_a through the textile layer

To obtain the final equation for the heat balance, the concept proposed by Burton and Edholm (1955) for F_{cl} and the coefficient of the permeability efficiency of the clothing F_{pcl} is applied. Both factors are similar and concern the convective heat transfer and mass transfer by water vapor, respectively; at the same time, they are a function of the clothing insulation I_{cl}, and the corresponding coefficients h and h_c.

The full equation for the heat balance of the body in a uniform environment, provided that the individual does not perform external work W, has the form

$$S_b = M[1 - 0.0023(44 - \phi_a P_a)]$$
$$-2.2h_c(0.06 + 0.095w_{rsw})(P_{sk} - \phi_a P_a)F_{pcl} - (h_r + h_c)(T_{sk} - T_a)F_{cl} \tag{16.12}$$

The first term of the equation reflects the heat generated by the body, which is lost through the skin surface, W/m²; the second—the total heat losses by evaporative heat transfer from the skin surface, W/m²; and the third—the dry heat exchange, W/m².

16.3.3.2 Heat Exchange between the Core Body and the Skin

The total mass of the body m_b is a sum of the mass of the skin m_{sk} and the mass of the core body m_{cr}, that is:

$$m_b = m_{sk} + m_{cr} \tag{16.13}$$

For a body weight of 81.7 kg, the core body weight is 78.3 kg, and the skin weight is 3.4 kg (Gagge et al. 1971).

The heat flow from and to the skin of the body can be calculated from the following expression:

$$S_{sk} = K_{min}\left(T_{cr} - T_{sk}\right) + c_{bl}\dot{V}_{bl}\left(T_{cr} - T_{sk}\right) - E_{sk} - (R + C) \tag{16.14}$$

where:

S_{sk} is the heat stored in the skin, W/m²

K_{min} is the minimum conductivity of the skin in the case of absence of blood flow, m²K

c_{bl} is blood's specific heat, $c_{bl} = 1.163$ kJ/kgK

$\dot{V}_{bl}$ is the rate of the blood flow in the skin, l/hm²

T_{cr} is the temperature of the core of the body, °C

T_{sk} is the skin temperature, °C

The heat flow from and to the core body is given by the relation

$$S_{cr} = \left(M - E_{res} - W\right) - K_{min}\left(T_{cr} - T_{sk}\right) - c_{bl}\dot{V}_{bl}\left(T_{cr} - T_{sk}\right) \tag{16.15}$$

The total heat S_b, stored by the body, is

$$S_b = S_{cr} + S_{sk}, \text{W/m}^2 \tag{16.16}$$

The total thermal capacity of the skin and the core body can be calculated as follows:

$$c'_{sk} = 0.97 m_{sk}, \text{W/°C} \quad \text{and}$$
$$c'_{cr} = 0.97 m_{cr}, \text{W/°C} \tag{16.17}$$

where the prime sign means that the value is valid for the whole skin or core body, and 0.97 is the specific heat of the body, W/kg°C

The rate of change of the skin temperature $\dot{T}_{sk}$ or core body temperature $\dot{T}_{cr}$ can be determined as

$$\dot{T}_{sk} = S_{sk}A/c'_{sk}, \text{°C/h} \quad \text{and}$$
$$\dot{T}_{cr} = S_{cr}A/c'_{cr}, \text{°C/h} \tag{16.18}$$

In the equations for the temperature change, Newtonian heating and cooling of the body (the core and the skin) is considered, that is, the temperature changing is uniform. The model also assumes that the DuBois surface of the body A (DuBois and DuBois 1916) is the same for the core and skin of the body. Provided that the temperature of the skin and the core body in the initial exposure of the body to a temperature T_a and relative humidity ϕ_a are 34.1°C and 36.6°C, respectively, at any time it can be written that

$$T_{sk} = 34.1 + \int_0^t \dot{T}_{sk} dt \quad \text{and}$$

(16.19)

$$T_{cr} = 36.6 + \int_0^t \dot{T}_{cr} dt$$

16.3.4 Control System

This part of the Gagge's model controls the blood flow from the core body to the skin and the latent heat exchange mechanisms (through sweating). It has been found that the average skin temperature is 34.1°C and the average temperature of the core body is 36.6° (mentioned already in Equation 16.19) in the event that the energy of the body for thermoregulation is minimal (Stolwijk and Hardy 1966). When both values are in the rest of the body, it is assumed that the person is in a state of *physiological thermal neutrality* (no reaction of the circulatory system or sweating).

On this basis, Gagge's model determine the equations for warm and cold signals:

$$\Sigma_{sk} = T_{sk} - 34.1 \quad \text{and}$$

$$\Sigma_{cr} = T_{cr} - 36.6$$

(16.20)

The reactions of the two layers of the model are as follows:

- Skin reactions
  - Cold sensation when $\Sigma_{sk} < 0$; it provokes vasoconstriction of the blood vessels in the skin and decreases the heat transfer from the core to the skin.
  - Warmth sensation when $\Sigma_{sk} > 0$; it provokes sweating that has greater importance for the thermoregulation of the body than the vasodilatation.
- Core body reactions
  - Cold sensation when $\Sigma_{cr} < 0$; it provokes vasoconstriction, which has lower rate and effectiveness than the vasoconstriction as a skin reaction.
  - Warmth sensation when $\Sigma_{cr} > 0$; it provokes vasodilatation and sweating.

The following has been established in the multisegment model of Stolwijk and Hardy (1966):

- *For the skin layer*: When the signals are $\Sigma_{sk} < 0$, the blood flow encounters proportional increment of the resistance for each °C temperature drop; on hands and feet, this resistance factor can increase twice for each °C decrease of the temperature of the environment; the vasoconstriction of the blood vessels in the torso is negligible.
- *For the core body layer*: When the signals are $\Sigma_{cr} > 0$ for each °C of temperature increment, the blood flow increases with 75 l/hm² over the *normal* blood flow of 6.3 l/hm²—a value that is valid for the case of body rest and thermal neutrality.

On the basis of these data, the rate of blood flow for the skin layer in the Gagge's model is calculated by

$$\dot{V}_{bl} = \frac{6.3 + 75\Sigma_{cr}}{1 - 0.5\Sigma_{sk}}, 1/hm^2 \tag{16.21}$$

In Equation 16.21, when $\Sigma_{cr} < 0$ (signal for *cold* as $T_{cr} < 36.6$) and/or $\Sigma_{sk} > 0$ (signal for *warmth* as $T_{sk} > 34.1$), the signal in both cases is considered to be equal to 0, that is, $\Sigma = 0$.

The glands that provoke perspiration on the surface of the skin layer to assure the thermoregulation by latent heat exchange are activated by signals from either the core body (Σ_{cr}) or the product $(\Sigma_{sk})(\Sigma_{cr})$. For the rate of sweating $\dot{m}_{rsw}$ is valid:

$$\dot{m}_{rsw} = 250\Sigma_{cr} + 100(\Sigma_{cr})(\Sigma_{sk}) \tag{16.22}$$

The first term of Equation 16.22 is important mainly for energetic activity. Satin et al. (1970) found that any increase in the temperature of the core (i.e., the basal temperature) over 36.6°C gives rise to the sweating with an average of 250 g/hm². The second term describes the sweating during rest (Hardy and Stolwijk 1966). The multiplier 100 reflects the dual effect of the signal received from the product $(\Sigma_{cr})(\Sigma_{sk})$, which, in turn, has a minor importance during activity, as T_{sk} decreases significantly below 34.1°C.

Bullard et al. (1970) demonstrated the effect of the skin temperature on the local change of the production of water vapor from the skin surface. This dimensionless ratio was included in the Gagge's model as a description of the heat losses from the latent heat exchange, namely:

$$E_{rsw} = 0.7\dot{m}_{rsw}2^{(T_{sk}-34.1)/3} \tag{16.23}$$

where the power function $2^{(T_{sk}-34.1)/3}$ is a dimensionless coefficient, which shows that the increment of T_{sk} with 3°C over 34.1°C facilitates twice the production of secretions from the sweat glands (i.e., in the presence of a source of heat radiation in the vicinity of the body). In an analogous manner, the decrement of T_{sk} with 3°C under 34.1°C decreases the production of sweat to a half. The factor 0.7 reflects the latent heat of sweat.

The Gagge's model was described in detail in Gagge et al. (1971), where the FORTRAN version of the software code was also presented.

16.4 Implementation of the Model in FLUENT CFD Software Package

The implementation of a thermophysiological model of a human body in a CFD software package that allows an accurate assessment of the environmental parameters is a modern way to assess the human thermophysiological comfort, as it leads to increased opportunities for simulating the impact of the dynamic environment on the human body, and the opposite. While essential part of nowadays thermophysiological models requires the environmental parameters to be set as input data, the joint simulation of the aerodynamics of the room and the comfort of the occupants allows the modeling of very close to reality situations in dynamic, interconnected mode.

It is not possible to accurately predict the effects of the environment on the thermoregulatory system of the human body without the knowledge of the aerodynamics of the ventilated room and the thermal environment. On the other hand, the assessment of the indoor environment parameters, that is, air exchange rate, and temperature, are not sufficient to evaluate the thermal comfort of the occupants, without simulating the thermal response of the human body.

The implementation of the Gagge thermophysiological model in FLUENT CFD software package was done by Pichurov (2009), and details are also presented in Pichurov (2008), Pichurov and Stankov (2013), and Pichurov et al. (2014).

16.5 Summary

The mechanisms of thermoregulation of the human body in terms of environmental parameters and simulation with thermophysiological model were systematized and analyzed.

Classifications of existing thermophysiological models were presented. The advantages of the thermophysiological model of Gagge and details of its passive and control system were detailed.

17

Numerical Study of the Thermophysiological Comfort of a Clothed Body

The main aim of this part of the study was to investigate the interaction between clothed human body and the indoor environment with its specific parameters through a layer(s) of textiles (clothing) with different insulating properties, when performing different activity, which—as a result—causes different reactions of the thermoregulation system of the human body.

17.1 Specifics of the Study

In particular indoor environment, where the thermophysiological comfort of the individual is determined by the heating, ventilation, and air conditioning systems, or their absence (when natural ventilation is applied), clothing and activity are the only two factors that can be controlled by the individual and can be *managed* within certain limits (Angelova 2010c). In many situations, however, the person performs routine, repetitive actions that do not lead to significant changes in metabolic reactions in cells, especially in muscle tissues. At the same time, many professions, exercised in the indoor environment, require special clothing that cannot be changed either because of the existence of a dress code (offices), uniforms (restaurants, hotels, airplanes), or because of the existence of protective clothing (hospitals, manufacturing facilities).

The numerical modeling and computer simulation of the thermophysiological reactions of the human body in a particular indoor environment with specific parameters and features is an up-to-date and complex task, which is interdisciplinary in its nature. The specific environmental conditions, as was mentioned in Chapter 1, significantly affect the human thermophysiological comfort, but at the same time the human body is an active source of heat, which is involved in the exchange of heat and moisture with the indoor environment (Pichurov et al. 2014). The modeling of this interaction by using a mathematical model to predict the response of the thermoregulation system of the human body in any activity and in any environment is a task that can be solved by thermophysiological models (detailed in Chapter 16). The integration of such a thermophysiological model in the simulation of room air flow improves the description of the generation and transfer of heat in and from the human body, which allows to obtain numerical results for core body temperature and skin temperature, as well as the heat losses (Pichurov and Stankov 2013).

The problem of the *human body–environment* interaction was a subject of numerous publications in recent decades. Some of them were studies based on computational fluid dynamics (CFD) as a powerful tool for predicting and analyzing the interdependence between the environment and the body reactions, including in the design stage of special clothing with required thermal insulation (Antunano and Nunneley 1992; Tan et al. 1998; Li 2007; Cheng et al. 2012; Epstein et al. 2013; Schellen et al. 2013). A series of studies were based on the use of thermal mannequins for assessment of the internal environment in buildings (Murakami et al. 2000; Al-Mogbel 2003; Søresen and Voigt 2003; Gao and Niu 2004; Sideroff and Dang 2005; Zhou et al. 2013). Many field studies on the effect of the environment on human comfort and health were also reported (Naydenov et al. 2006; Carvalho et al. 2013; Fabbri 2013; Ivanov et al. 2014; Markov et al. 2014). The implementation of the thermophysiological model in a computational procedure for calculating the parameters of the indoor environment is a very modern approach for simulating the thermal sensation of the interaction between the individual and the thermoregulation mechanisms of the body at specific environmental conditions (Pichurov 2009; Van Treeck et al. 2009; Fiala et al. 2010; Cheng et al. 2012; Pichurov and Stankov 2013).

The last approach was used in this study, having as a background the implementation of the Gagge model in FLUENT CFD software package, performed by Pichurov (2009). Details of the investigation were presented in Pichurov et al. (2014) and Angelova et al. (2015).

It should especially be noted that the original model of Gagge (Gagge et al. 1971) is valid integrally for the whole body, while with its implementation in FLUENT CFD software package it is applied locally for each computational cell of the body surface (Pichurov and Stankov 2013; Pichurov et al. 2014). In addition, a layer of clothing is added to the Gagge's model, which complicates the task, but makes it much closer to the real physical problem as it reflects the complex interaction *human body–textiles–indoor environment* (Angelova et al. 2015).

17.2 Mathematical Model

To solve the problem of simulating the thermophysiological comfort of a clothed body in real ventilated room with aerodynamic interaction, a mathematical model based on Reynolds-averaged Navier–Stokes equations (RANS) was used.

The Reynolds system of partial differential equations was presented in Section III. The continuity equation and the equation for heat transfer in turbulent flows were added to the system. A standard $k - \varepsilon$ turbulent model was used; thus, two additional transport equations were added to the RANS system: for the kinetic energy of the turbulent pulsations k and its dissipation rate ε.

The final system of PDEs closed with the $k - \varepsilon$ turbulent model is

$$\frac{\partial u}{\partial t} + u\frac{\partial u}{\partial x} + v\frac{\partial u}{\partial y} + w\frac{\partial u}{\partial z} = X - \frac{1}{\rho}\frac{\partial p}{\partial x} + v\left(\frac{\partial^2 u}{\partial x^2} + \frac{\partial^2 u}{\partial y^2} + \frac{\partial^2 u}{\partial z^2}\right)$$

$$-\rho\left(\frac{\partial \overline{u'}^2}{\partial x} + \frac{\partial \overline{u'v'}}{\partial y} + \frac{\partial \overline{u'w'}}{\partial z}\right)$$

(17.1)

$$\frac{\partial v}{\partial t} + u\frac{\partial v}{\partial x} + v\frac{\partial v}{\partial y} + w\frac{\partial v}{\partial z} = Y - \frac{1}{\rho}\frac{\partial p}{\partial y} + v\left(\frac{\partial^2 v}{\partial x^2} + \frac{\partial^2 v}{\partial y^2} + \frac{\partial^2 v}{\partial z^2}\right)$$

$$-\rho\left(\frac{\partial \overline{u'v'}}{\partial x} + \frac{\partial \overline{v'}^2}{\partial y} + \frac{\partial \overline{v'w'}}{\partial z}\right)$$

(17.2)

$$\frac{\partial w}{\partial t} + u\frac{\partial w}{\partial x} + v\frac{\partial w}{\partial y} + w\frac{\partial w}{\partial z} = Z - \frac{1}{\rho}\frac{\partial p}{\partial z} + v\left(\frac{\partial^2 w}{\partial x^2} + \frac{\partial^2 w}{\partial y^2} + \frac{\partial^2 w}{\partial z^2}\right)$$

$$-\rho\left(\frac{\partial \overline{u'w'}}{\partial x} + \frac{\partial \overline{v'w'}}{\partial y} + \frac{\partial \overline{w'}^2}{\partial z}\right)$$

(17.3)

$$\frac{\partial u}{\partial x} + \frac{\partial v}{\partial y} + \frac{\partial w}{\partial z} = 0$$

(17.4)

$$\frac{\partial}{\partial x}\left[\rho u k - \left(\mu + \frac{\mu_t}{\sigma_k}\right)\frac{\partial k}{\partial x}\right] + \frac{\partial}{\partial y}\left[\rho v k - \left(\mu + \frac{\mu_t}{\sigma_k}\right)\frac{\partial k}{\partial y}\right]$$

$$+ \frac{\partial}{\partial z}\left[\rho w k - \left(\mu + \frac{\mu_t}{\sigma_k}\right)\frac{\partial k}{\partial z}\right] = P - \rho\varepsilon$$

(17.5)

$$\frac{\partial}{\partial x}\left[\rho u \varepsilon - \left(\mu + \frac{\mu_t}{\sigma_\varepsilon}\right)\frac{\partial \varepsilon}{\partial x}\right] + \frac{\partial}{\partial y}\left[\rho v \varepsilon - \left(\mu + \frac{\mu_t}{\sigma_\varepsilon}\right)\frac{\partial \varepsilon}{\partial y}\right]$$

$$+ \frac{\partial}{\partial z}\left[\rho w \varepsilon - \left(\mu + \frac{\mu_t}{\sigma_\varepsilon}\right)\frac{\partial \varepsilon}{\partial z}\right] = \frac{\varepsilon}{k}(c_1 G - c_2 \rho \varepsilon)$$

(17.6)

$$G = \mu_t \left\{ \begin{array}{l} 2\left[\left(\dfrac{\partial u}{\partial x}\right)^2 + \left(\dfrac{\partial v}{\partial u}\right)^2 + \left(\dfrac{\partial w}{\partial z}\right)^2\right] + \left(\dfrac{\partial u}{\partial y} + \dfrac{\partial v}{\partial x}\right)^2 + \left(\dfrac{\partial v}{\partial z} + \dfrac{\partial w}{\partial y}\right)^2 \\[4mm] + \left(\dfrac{\partial w}{\partial x} + \dfrac{\partial u}{\partial z}\right)^2 \end{array} \right\} \tag{17.7}$$

$$\mu_{\text{eff}} = \mu + \mu_t \tag{17.8}$$

$$\mu_t = c_\mu \rho \frac{k^2}{\varepsilon}$$

$$\frac{\partial}{\partial x}\left[\rho u T - \left(\frac{\mu}{\sigma} + \frac{\mu_t}{\sigma_t}\right)\frac{\partial T}{\partial x}\right] + \frac{\partial}{\partial y}\left[\rho v T - \left(\frac{\mu}{\sigma} + \frac{\mu_t}{\sigma_t}\right)\frac{\partial T}{\partial y}\right]$$

$$+ \frac{\partial}{\partial z}\left[\rho w T - \left(\frac{\mu}{\sigma} + \frac{\mu_t}{\sigma_t}\right)\frac{\partial T}{\partial z}\right] = 0 \tag{17.9}$$

Standard values of the model constants were used, namely, $c_\mu = 0.09$, $\sigma_k = 1$, $\sigma_\varepsilon = 1.3$, $C_1 = 1.44$, and $C_2 = 1.92$.

17.3 Procedure of the Study

To evaluate the thermophysiological comfort of the human body in its complexity in a random indoor environment, this part of the study aimed to assess the impact of the three major components that preconditioned this comfort:

- The impact of the indoor environment (the room)—by simulating different aerodynamics and temperature patterns.
- The impact of the textile barrier between the body and the environment—by changing the thermal insulation properties of the textile layer.
- The impact of the reactions of the body (thermoregulation system)— by changing the metabolic activity.

17.3.1 Local Thermophysical Properties of Textiles and Clothing

Several studies (Li and Holcombe 1998; Havenith 2001; Pichurov 2009) commented on the need for precise knowledge of the thermal insulation properties of textiles as the numerical results, based on the application of thermophysiological models, depend directly on them. At the same time, the authors noted that the task is not trivial and requires an interdisciplinary approach.

In most cases, both authors and users of thermophysiological models as well as researchers, who deal with experimental assessment of the thermo-physiological comfort using thermal manikins or field measurements with real individuals, are using the databases of thermophysical parameters of clothing, included in ASHRAE (1993) and ISO 7730 (1995). The two sources, together with publications, like these of McCullough et al. (1985, 1989), provide information on thermal resistance, air and vapor permeability of sets of clothes, rarely of single items. Most of the results are obtained by studies with a thermal manikin in climate chambers. The disadvantage of this method of measurement is that it determines the effect of the item or the clothing ensemble on the total power consumption of the mannequin (Pichurov 2008). On the other hand, the use of thermal mannequin mimics in the utmost degree the presence of air layers between the layers of clothing, as a function of the shape of the body and the posture at that, which cannot be measured with the apparatus, based on disk method, for example.

17.3.2 Selection of Clothing

Five values of the thermal insulation properties of the textile layer over the skin of the thermophysiological model were used (ASHRAE Handbook 1997).

- *0.37 clo*: Shorts, T-shirt with short sleeves
- *0.60 clo*: Trousers, T-shirt with long sleeves
- *0.93 clo*: Trousers, long sleeve shirt and jacket
- *1.3 clo*: Padded trousers, T-shirt, long sleeve shirt, sweater, jacket
- *1.53 clo*: Padded trousers, T-shirt, long sleeve shirt, sweater, padded jacket

Similarly, combinations of textiles whose insulating properties were already known could be selected, as some of the textiles, which were subject to experimental study presented in Section II.

17.3.3 Selection of Activity

Three types of activities, performed in the indoor environment, were selected (Tudor-Locke et al. 2011):

- *1.2 met (70 W/m²)*: Activity without movement, most often sedentary (office work, waiting in line, banking/financial services, police and medical services, mental activity at rest, listening to music, smoking at rest and others)

- *2.0 met (116 W/m²)*: Moderate activity (care for children or elderly persons at home, general housekeeping, administration, training, teaching, repair of clothes, cleaning, preparing meals at home and others)
- *3.0 met (174 W/m²)*: Active work (care for pets, bowling, training weightlifting, or bodybuilding, housekeeping, ironing, and others)

17.3.4 Parameters of the Indoor Environment

Two general situations were studied:

- *Summer cases with cooling*: It corresponds to summer conditions with cooling from the air-conditioning system, that is, the supply air temperature was set to 22°C and velocity of 2 m/s. The supply opening

TABLE 17.1

Simulated Cases

Case	Indoor Environment	Clothing Thermal Resistance R_{cl}, m²K/W	Clothing Insulation I_{cl}, clo	Activity W/m²	met
1	Summer, cooling	0.057	0.37	70	1.2
2	22°C			116	2.3
3				174	3.0
4		0.093	0.60	70	1.2
5				116	2.3
6				174	3.0
7		0.144	0.93	70	1.2
8				116	2.3
9				174	3.0
10		0.200	1.3	70	1.2
11				116	2.3
12				174	3.0
13		0.237	1.53	70	1.2
14				116	2.3
15				174	3.0
16	Summer, no	0.057	0.37	70	1.2
17	cooling 28°C			116	2.3
18				174	3.0
19		0.144	0.93	70	1.2
20				116	2.3
21				174	3.0
22		0.200	1.3	70	1.2
23				116	2.3
24				174	3.0

was on the top left corner, while the exhaust opening was on the bottom right corner.

- *Summer cases without cooling*: The temperature of the supplied air was set to 28°C.

17.3.5 Parameters of the Human Body

The thermophysiological model integrated with FLUENT 6.3 CFD software package (Pichurov 2009; Pichurov and Stankov 2013) had a body height of 1.65 m. The total body contour was 3.3 m and the DuBois body area A was 1.89 m². The clothing area factor was $f_{cl} = 1.2$ (Pichurov et al. 2014).

17.3.6 Summary of the Investigated Cases

The conditions of the investigated cases were summarized in Table 17.1 (Pichurov et al. 2014; Angelova et al. 2015). Structured grid was generated for all cases.

17.4 Summary

The essence of the computational task for simulation of the interaction between the individual and the thermoregulatory mechanisms of the body in specific environmental conditions, based on Gagge thermophysiological model, was presented. The simulated cases were also presented, detailing the choices of clothing insulation levels and activities in the indoor environment.

18

Results from the Numerical Study on the Thermophysiological Comfort of a Clothed Body

This chapter presents sequentially numerical results that allow to assess the impact of the three main factors that determine the thermophysiological comfort of a clothed person (Pichurov et al. 2014; Angelova et al. 2015):

- Effects of the environment (room air flow and temperature distribution)
- Influence of clothing thermal insulation (insulation of textile layer or layers)
- Effects of the activity of the individual

18.1 Influence of the Environmental Parameters

Stankov (1998) demonstrated by numerous examples that the modeling of the process of ventilation even in rooms with simple geometry (rectangular geometry, no furniture and inhabitants) is very complex. In the present study, the selected room had a simplified geometry, but even in this case, the complexity of the aerodynamics of the room was obvious (Pichurov et al. 2014).

Due to the impact of colder air, entering the room from the opening in the upper left corner, two temperature zones appeared in the room, as seen from the isotherms for case 8 (0.93 clo, 2.3 met) (Figure 18.1).

The cold air flow was sticking to the ceiling and moved down to the floor on the left side of the body, creating a zone with a lower temperature, compared to the right side of the body. The appearance of two zones with different air temperature would precondition different heat losses from the two sides of the body. The results confirmed the conclusion, drawn in Stankov (1998), that the complexity of the aerodynamic interaction in enclosures was determined largely by the scheme of the air supply.

Obviously, the knowledge on the local parameters of the flow in the ventilated space is essential for the assessment of the thermophysiological comfort

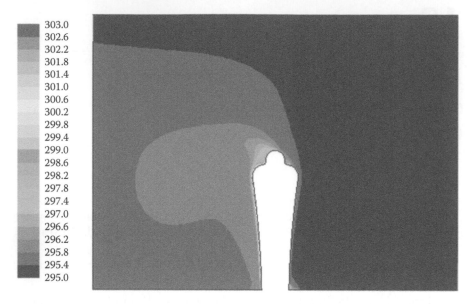

FIGURE 18.1
Air temperature distribution, K: case 8 (0.93 clo, 2.3 met).

of the occupants. In this case, the performed numerical modeling permits to obtain a database for the temperature and local velocity components in any point of the room.

To assess the impact of the two temperatures of the supplied air in the room, a comparison between the isotherms for selected similar cases was made.

Figures 18.2 and 18.3 show the temperature distribution for cases 9 and 21 (0.93 clo, 3.0 met), with and without cooling. There was a clear difference in the profiles of the contours and the temperature. On the right side of the body, for example, the temperature of the room air was within the range of 21.85°C–22.35°C in case 9 (cooling), while in case 21 the temperature range was 27.85°C–28.35°C.

The temperature asymmetry in the room affected the results for the surface temperature of both clothing T_{cl} and skin T_{sk}, since the room temperature T_a was one of the factors that influenced the heat exchange of the system *body–textiles–indoor environment*. This was clearly shown in Figure 18.4: the surface temperature of the skin T_{sk} on the right side of the body was higher than that on the left side. That result was valid for all tested cases.

As should be expected, the temperature of the skin surface was higher in the cases with no cooling of the room. The average value of T_{sk} for case 1 is 33.88°C, while in the corresponding case 16 it was 35.4°C (Figure 18.4). It is noteworthy, however, the more even temperature changes around the body contour in the case without cooling (case 16), while in case 1 (with cooling)

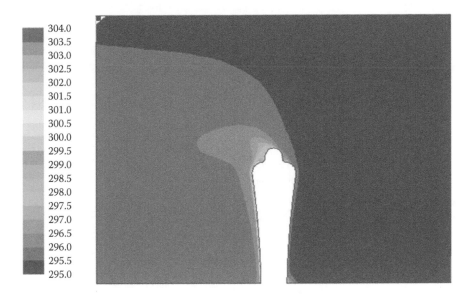

FIGURE 18.2
Isotherms, K: case 9, cooling (22°C, 0.93 clo, 3.0 met).

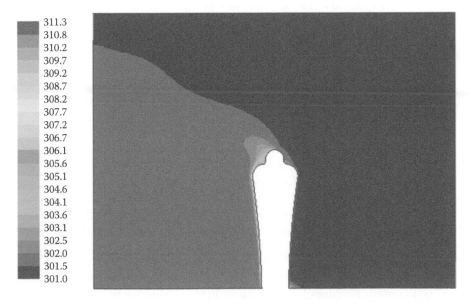

FIGURE 18.3
Isotherms, K: case 21, no cooling (28°C, 0.93 clo, 3.0 met).

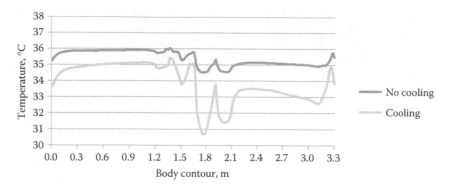

FIGURE 18.4
Skin temperature, °C: cases 1 (cooling) and 16 (no cooling), 0.37 clo, 1.2 met.

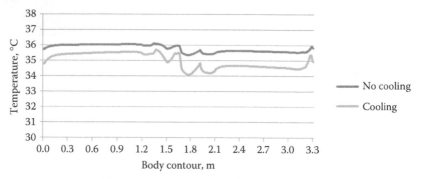

FIGURE 18.5
Skin temperature, °C: cases 7 (cooling) and 19 (no cooling), 0.37 clo, 1.2 met.

the temperature asymmetry between the right and left side of the body was more noticeable.

The results in Figures 18.4 and 18.5 allowed a comparison between the skin temperature to be made when different clothing insulation was applied (0.37 and 0.93 clo) for one and the same activity (1.2 met). For ease of reading, the ordinates of Figures 18.4 and 18.5 had the same scales. Obviously, the higher thermal insulation of the garment (Figure 18.5) provoked an increase in T_{sk} in both cases with and without cooling of the room, due to more difficult heat exchange between the body and the ambient air. The increment of the average skin temperature for the case of cooling was 3.1%, and for the case without cooling—1%.

Figure 18.6 shows typical result for the clothing temperature changes around the body contour for cases 1 (cooling) and 16 (no cooling). Obviously, the overall temperature of the surface of the clothing was lower than the skin temperature (Figure 18.4) and that result was valid for all tested case. Moreover, as expected, T_{cl} was higher for the cases without cooling of the room air.

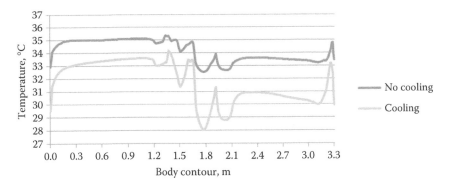

FIGURE 18.6
Clothing temperature, °C: cases 1 (cooling) and 16 (no cooling), 0.37 clo, 1.2 met.

The temperature asymmetry on the left and right side of the body was obvious for T_{cl} as well, and this result was again similar for all cases. Obviously, the impact of clothing insulation and metabolism could be again analyzed, by analogy with the analysis of the results for T_{sk}.

18.2 Influence of the Clothing Insulation

The presence of a textile layer(s) between the human skin and the environment, whether in the form of clothing or blanket, leads to alteration of the heat exchange between the body and the indoor environmental air, as was shown in Chapter 2.

In the conditions of thermal neutrality, the skin temperature on the torso is around 33°C, and that of the skin of the feet is approximately 30°C–31°C (Angelova 2007). In this case, the individual is in a state of thermophysiological comfort, as the mean skin temperature of 33°C corresponds to an average value of the core temperature of 37°C (Havenith 2003).

Figures 18.7 and 18.8 show the results from the simulation of the effect of the clothing thermal insulation on the skin temperature T_{sk} for two of the tested values of metabolic activity.

The temperature around the body contour in Figure 18.7 for sedentary activity (1.2 met) showed that the impact of I_{cl} is much noticeable at the left side of the body (1.65–3.3 m) than on the right (0–1.65 m). The reason for this was that the temperature difference between the air temperature T_a and skin temperature T_{sk} was greater on the left side of the body.

While the thermal insulation of the clothing managed to maintain T_{sk} on the right side of the body between 34°C (for 0.37 clo) and 35.8°C (for 1.54 clo), on the left side of the body T_{sk} varied from 31.5°C (for 0.37 clo) to 35.2°C (for

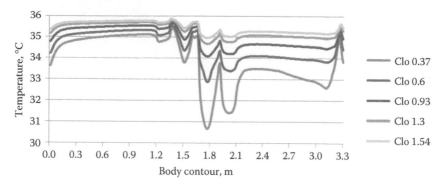

FIGURE 18.7
Skin temperature, °C, activity 1.2 met.

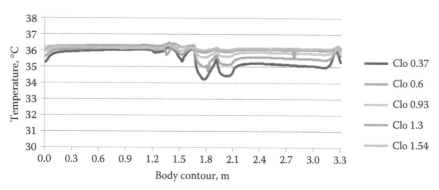

FIGURE 18.8
Skin temperature, °C, activity 2.3 met.

1.54 clo). Thus, a real occupant would feel thermal discomfort due to the temperature asymmetry in real conditions. Despite the visualized temperatures, however, the mean skin temperature for the discussed cases (cases 1, 4, 7, 10, and 13) in Figure 18.7 was changing smoothly and continuously increased with the augmentation of the clothing insulation I_{cl}: $\overline{T}_{sk} = 33.88$, 34.53, 34.98, 35.30, 35.45°C for the respective cases.

The increment of the metabolic activity from 1.2 (Figure 18.7) to 2.3 met (Figure 18.8) caused more uniform change of T_{sk} on the right and left side of the body. The skin temperature on the right side of the body (Figure 18.8) was around 36°C for all values of I_{cl}, while on the left side of the body it changed with 1°C–1.5°C between the five cases (cases 2, 5, 8, 11, and 14).

Figures 18.9 and 18.10 visualize the results for the temperature on the clothing surface T_{cl} as a function of the clothing insulation I_{cl}, comparing again different levels of the metabolic activity.

In contrast to the results for the skin temperature, the graphs of the temperature of the clothing surface had one and the same tendency, without a

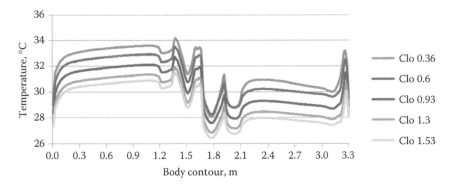

FIGURE 18.9
Clothing temperature, °C, activity 1.2 met.

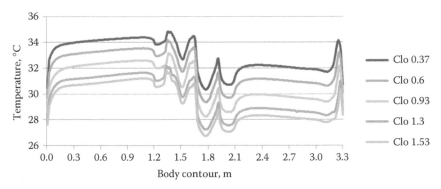

FIGURE 18.10
Clothing temperature, °C, activity 2.3 met.

marked effect of the activity level. The higher values of I_{cl} provoked lower values of T_{cl} due to the smaller intensity of the heat transfer through the textile barrier. For example, the decrement of T_{cl} for a level of 1.0 m of the body contour on Figure 18.9 was 2%–3% (for 0.6 clo) and 9.6% (for 1.53 clo) in comparison with the lowest value of the thermal insulation of 0.36 clo. With an increase in the activity (Figure 18.10), the clothing temperature for the same point of the body contour changed to 4.1% (for 0.6 clo) and 11.9% (for 1.53 clo).

Figure 18.11 presents a comparison between skin and clothing temperature for case 8 (clothing insulation 0.93 clo, activity 2.3 met).

There was a much smaller dependence of the T_{sk} on the temperature asymmetry in the room as compared to T_{cl} along the contour of the body: the thermoregulatory system of the human body did not allow significant changes in temperature, especially in core body, regardless of the environmental conditions. Excess heat would be removed by any of the mechanisms for thermoregulation, most often by the latent heat exchange.

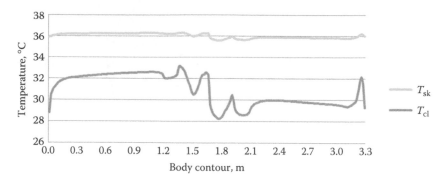

FIGURE 18.11
Skin and clothing temperature, °C, 0.93 clo, 2.3 met.

18.3 Influence of the Activity

In the conditions of performed activity, the body leaves the comfort zone of thermal neutrality and generates heat, due to the metabolic processes in each cell of the body. As a result, both skin temperature and core body temperature increase.

Figure 18.12 summarizes the numerical results for the skin temperature T_{sk} as a function of the performed activity for a constant clothing insulation $I_{cl} = 0.6$ clo (cases 4–6). The graphs showed the effect of the temperature asymmetry in the room for the activity 1.2–2.3 met. By increasing the activity to 3.0 met, T_{sk} became more uniform along the entire contour of the body and ceased to reflect the asymmetry of the temperature environment.

The same was valid for the results shown in Figure 18.13 for a higher clothing insulation $I_{cl} = 0.93$ clo (cases 7–9) and the same levels of activity.

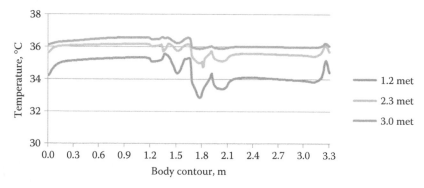

FIGURE 18.12
Skin temperature, °C, 0.6 clo.

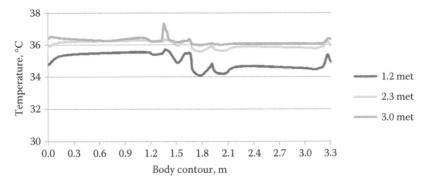

FIGURE 18.13
Skin temperature, °C, 0.93 clo.

However, even the increment of the activity to 2.3 met ceased in that case the influence of the temperature asymmetry in the room.

The graphs for T_{sk} showed the expected increase in temperature with the increment of the activity, while maintaining a constant insulation of clothing. The results showed clearly that this augmentation was not linear: the increase in T_{sk} was greater from 1.2 to 1.6 met, than from 1.6 to 2.3 met.

Another important result was that with the increase in the clothing insulation I_{cl}, the skin temperature T_{sk} became less dependent on the temperature of the indoor environment, due to the greater insulation barrier between the body and the air. That was an evidence of the creation of the so-called micro-environment by clothing (textile layer/layers) around the body whose role and importance were discussed in Section I.

As for the temperature on the surface of the clothing, it obviously followed the changes of the skin temperature, as it is shown in the exemplary Figure 18.14 for clothing insulation 0.6 clo.

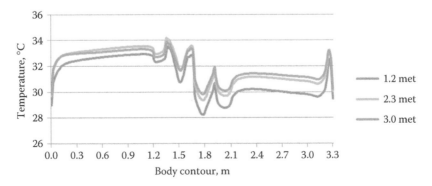

FIGURE 18.14
Clothing temperature, °C, 0.6 clo.

18.4 Application Analyses of the Numerical Results

The computational study on any real problem brings a series of requirements, the most important of which are as follows:

- Correct description of the physical phenomenon
- Application of a verified code (software)
- Obtaining of physically reliable numerical results (which is not
  necessarily fulfilled even the first two requirements are completed)
- Comparative costs of the study (as it concerns time, human resources,
  financial resources)
- Adequate application of the results in terms of processing, analysis,
  and feasibility

Here some additional exemplary analyzes, based on the numerical results for
skin and clothing temperatures, obtained in Pichurov et al. (2014) and Angelova
et al. (2015), are presented. The performed analyzes have engineering application and parts of them have been published in Angelova et al. (2015).

18.4.1 Skin Temperature and Clothing Temperature Isotherms

Figures 18.15 through 18.18 present the isotherms of the skin and clothing
temperature at 1.1 m height from the floor. This is typical plane for evaluation of the thermal comfort of a person in the indoor environment (ASHRAE
Handbook 2005).

The change is shown for every 1°C for all values of the thermal insulation
of clothing I_{cl} and three levels of activity. The graphs present the results on
the right and left sides of the body, which are different due to the temperature asymmetry in the room.

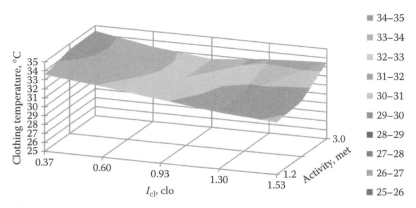

FIGURE 18.15
Isotherms for the clothing surface: 1.1 m from the floor, right-hand side of the body.

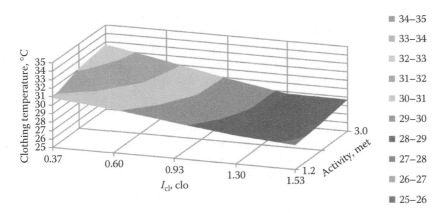

FIGURE 18.16
Isotherms for the clothing surface: 1.1 m from the floor, left-hand side of the body.

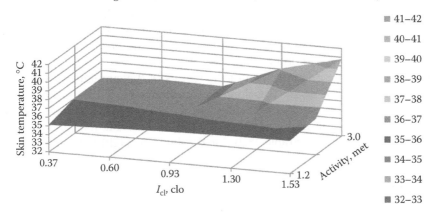

FIGURE 18.17
Isotherms for the skin surface: 1.1 m from the floor, right-hand side of the body.

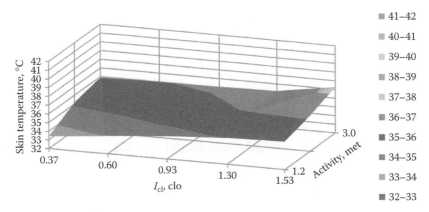

FIGURE 18.18
Isotherms for the skin surface: 1.1 m from the floor, left-hand side of the body.

The impact of the clothing isolation I_{cl} was obvious on both sides of the body (Figures 18.15 and 18.16). The decrement of the clothing temperature with the impediment of the heat transfer from the skin to the environment was more visible on the left side of the body (Figure 18.15) compared with the right side (Figure 18.16). The increase in activity also led to an increase in clothing temperature.

Similar is the change in skin temperature as a function of the increase in the activity level and the thermal insulation of the clothing I_{cl} (Figures 18.17 and 18.18). There is, however, a more intense temperature change on the right side of the body (Figure 18.17) for metabolic level above 2.3 met, when the thermal insulation of the garment is over 0.93 clo. The result illustrated a common situation in the indoor environment when the occupant felt undesirable state of local discomfort due to a difference in temperature of parts of the body: back–chest, left–right, and so on.

The temperature contours allow a quick visual assessment of the combinations of metabolic activities and thermal insulation of clothing, which will ensure the comfortable feeling of the individual in the living environment, to be done. In Figures 18.17 and 18.18, these were the zones with temperature in the range 36°C–37°C and below 36°C.

For more accurate quantification of the results of the simulation, additional post-processing of the results was done, as summarized in Tables 18.1 and 18.2.

The changes in temperature are shown as a percentage in comparison with the lowest level of thermal insulation of clothing I_{cl} (Table 18.1) and the lowest activity of the body (Table 18.2), for the left and right side of the body,

TABLE 18.1

Temperature Changes in Percentage, in Comparison with the Lowest Level of Clothing Thermal Insulation

Temperature	Body Side	Activity, met I_{cl}, clo	Percentage of Change in Comparison with the Lowest $I_{cl} = 0.37$ clo			
			0.60	0.93	1.30	1.53
Clothing	Right side	1.2	−1.98	−4.40	−6.70	−8.06
		2.3	−2.44	−5.18	−7.94	−9.23
		3.0	−3.24	−5.34	−3.50	−2.64
	Left side	1.2	−2.35	−5.34	−8.05	−9.56
		2.3	−3.32	−7.04	−10.42	−12.15
		3.0	−4.11	−8.42	−11.90	−12.08
Skin	Right side	1.2	0.61	1.21	1.64	1.85
		2.3	0.29	0.57	0.71	0.22
		3.0	1.18	0.87	8.24	12.86
	Left side	1.2	1.87	3.61	4.87	5.45
		2.3	1.11	1.99	2.53	2.78
		3.0	0.25	0.49	0.83	3.85

TABLE 18.2

Temperature Changes in Percentage, in Comparison with the Lowest Level of Activity

Temperature	Body Side	Activity, met I_{cl}, clo	Percentage of Change in Comparison with the Lowest Activity of 70 W/m²				
			0.60	0.93	1.30	1.53	0.60
Clothing	Right side	2.3	2.35	1.87	1.51	0.98	1.04
		3.0	2.48	1.16	1.47	5.99	8.52
	Left side	2.3	4.18	3.13	2.31	1.49	1.20
		3.0	5.97	4.06	2.52	1.53	3.02
Skin	Right side	2.3	2.67	2.33	2.02	1.73	1.03
		3.0	2.88	3.46	2.53	9.55	14.00
	Left side	2.3	5.16	4.38	3.54	2.83	2.51
		3.0	7.39	5.68	4.16	3.26	5.76

respectively, for both skin and clothing temperatures. The results again were presented for a plane at 1.1 m height above the floor.

The analysis of Table 18.1 showed that the temperature of the clothing surface from both left and right side of the body decreased more quickly than the skin temperature increased with the augmentation of the clothing thermal insulation I_{cl}. This decrement reached 8%–12%, while the temperature of the skin increased with 2%–5% for the individual cases. An exception were the cases, which were a combination between the highest levels of clothing insulation and activity, for the right side of the body, where the air temperature was also higher.

The same cases (1.3 and 1.53 clo, 3 met) resulted to be sensitive to the temperature change as a function of activity, calculated as a percentage change relative to the lowest activity (sedentary activity, 1.2 met) (Table 18.2). While in the other cases both the temperatures of the clothing and the skin changed in the range of 1%–5%, the combination of high thermal insulation of clothing and energetic activity apparently provoked a significant increase in temperature, which would result in thermal discomfort for the person.

18.4.2 Regression Analysis

The regression analysis of the numerical results, presented in Angelova (2015), is used for one more demonstration of the engineering feasibility of the results obtained.

18.4.2.1 Influence of the clo Value

The regression analysis allowed to find out the relationship between the temperature of the skin or the clothing layer and the clo values for different metabolic rates. Figure 18.19 presents as an example the numerical results for the skin temperature at 1 m above the floor, right-hand side of the body. The

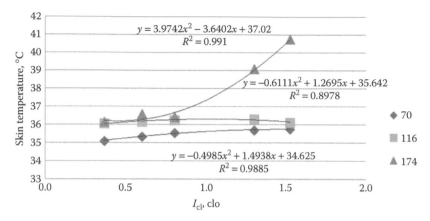

FIGURE 18.19
Skin temperature—right-hand side of the body: clothing insulation versus temperature.

respective trend lines, together with the determination coefficient R^2 for each line, were derived. The regression model for all activity levels was a second-level polynomial.

The same analysis could be performed for all numerical results for a particular plane of the body, that is, for right- and left-hand side, for both skin and clothing temperature. Table 18.3 summarizes the regression coefficients and the R^2 values for each regression model.

TABLE 18.3

Summary of the Regression Coefficients for Skin and Clothing Temperature Regression Models: Clothing Insulation versus Temperature

		Regression Model $y = ax^2 + bx + c$				
	Values	Activity, W/m²	a	b	c	R^2
Skin temperature	Right hand	70	−0.4985	1.4938	34,625	0.9885
		116	−0.6111	1.2695	35.642	0.8978
		174	−3.9742	−3.6402	37.020	0.9910
	Left hand	70	−1.3985	4.1332	32.163	0.9895
		116	−0.8807	2.4382	34.436	0.9865
		174	1.7552	−2.3599	36.664	0.8364
Clothing temperature	Right hand	70	1.1357	−4.4537	35.072	0.9906
		116	1.5537	−5.6678	36.260	0.9935
		174	4.3384	−8.6511	36.963	0.9106
	Left hand	70	1.3329	−5.0447	32.630	0.9899
		116	2.0947	−7.2833	34.639	0.9913
		174	3.7305	−10.495	36.184	0.9942

18.4.2.2 Influence of the Activity

The regression analysis could be used in the same way to find out the relationship between the temperature of the skin or the clothing layer and the metabolic rates (expressed by the met value) for different levels of clothing insulation. Figure 18.20 shows as an example the numerical results for the clothing temperature at 1 m, left-hand side of the body. The respective trend lines, together with the determination coefficient R^2 for each line, were derived. It was observed that a polynomial function of a third level best described the obtained results.

Following the same approach, regression analysis was applied to the numerical results so as to find out the determination coefficients for each case (right- and left-hand side, for both skin and clothing temperature at 1 m from the floor). Table 18.4 summarizes the regression coefficients and the R^2 values for each regression model.

18.4.2.3 Application of the Regression Models and Their Verification

To obtain results for the temperature of the skin or clothing for cases that were not subject of the performed numerical simulation, regression equations could be used. For example, to predict the skin temperature at a height of 1 m from the floor at the right-hand side of the body, further interpretation of the coefficients a, b, and c from the summarized results in Table 18.3 was performed. Figure 18.21 illustrates the points obtained and the regression equations, which were again polynomials of the second order. The results obtained for the coefficients k, l, and m are summarized in Table 18.5.

Following this principle, a series of regression equations can be derived and even nomograms can be built, if the task presumes a kind of a static aerodynamic picture of the room. Such an approach will facilitate the rapid

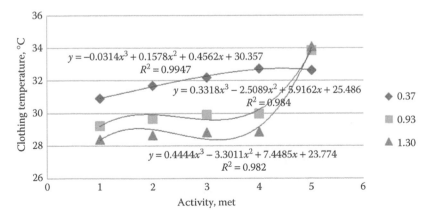

FIGURE 18.20
Clothing temperature—right-hand side of the body: activity level versus temperature.

TABLE 18.4

Summary of the Regression Coefficients for Skin and Clothing Temperature Regression Models: Activity Level versus Temperature

			Regression Model $y = ax^3 + bx^2 + cx + d$				
	Values	**Clo**	a	b	c	d	R^2
Skin temperature	Right hand	0.37	0.3546	−2.7196	6.5896	30.823	0.9846
		0.93	0.8948	−6.5925	14.718	26.366	0.9837
		1.30	0.7468	−4.9274	10.170	29.643	0.9987
	Left hand	0.37	−0.0440	0.2323	0.5402	32.780	0.9954
		0.93	0.5939	−4.4727	10.525	27.921	0.9854
		1.30	0.9235	−6.8463	15.473	25.366	0.9834
Clothing temperature	Right hand	0.37	0.2881	−2.2200	5.4109	30.043	0.9846
		0.93	0.6013	−4.4597	9.999	25.845	0.9811
		1.30	0.4628	−3.1148	6.5204	27.417	0.9988
	Left hand	0.37	−0.0314	0.1578	0.4526	30.357	0.9947
		0.93	0.3318	−2.5089	5.9162	25.486	0.9840
		1.30	0.4444	−3.3011	7.4485	23.774	0.9820

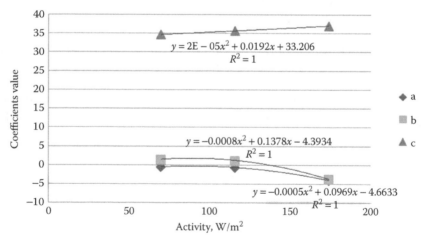

FIGURE 18.21

Skin temperature—right-hand side of the body: coefficients values.

determination of the temperature on the surface of the skin and clothing for different activity levels and clothing insulation.

As an example, the following task is shown: to assess the effect on the thermophysiological comfort of an employee, who, instead of a static desk job, will increase his activity, performing laboratory work in the same room, which has already being simulated. This means to determine the temperature of

TABLE 18.5

Summary of the Regression Coefficients k, l, and m for Skin Temperature, Right-Hand Regression Models: Activity Level versus Temperature

		$y = kx^2 + lx + m$			
		k	l	m	R^2
Skin temperature	a	0.0005	0.0969	−4.6633	1
Right hand	b	−0.008	0.1378	−4.3934	1
	c	0.00002	0.0192	33.206	1

the skin surface for an activity of 93 W/m² (1.6 met). As a result of the simulation, it is known that for the plane (1 m above the floor, right-hand side of body) the temperature on the skin surface is $T_{sk} = 35.18°C$, but when working in a seated position (70 W/m² or 1.2 met).

Provided that the clothing of the employee does not change, the regression equations allow to determine T_{sk} without applying new simulation. After calculating the coefficients a, b, and c, the regression equation has the form

$$T_{sk} = 0.0239I_{cl}^2 + 1.5028I_{cl} + 35.1646 \tag{18.1}$$

After substituting the specific value for the clothing insulation of I_{cl}, in this case $I_{cl} = 0.37$ clo, it is calculated that $T_{sk} = 35.72°C$, which shows an increase of 1.5% compared to static desk job.

Figure 18.22 shows a comparison between results, taken from the above mentioned calculations, and real numerical results for the same regime, taken from computational fluid dynamics (CFD) simulation. The results show very good coincidence, as the relative error is 0.005% for 0.37 clo, 1.62% for 0.93 clo, and 2.6% for 1.3 clo.

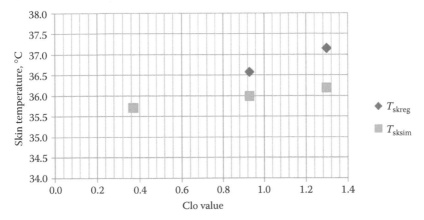

FIGURE 18.22
Skin temperature—results from a nomogram and a CFD simulation.

18.5 Summary

The effect of metabolic activity on the surface temperature of the human body and the outermost layer clothing (at constant insulation of clothing) was studied, together with the effect of the thermal insulation of clothing I_{cl} on the surface temperature of the human body and clothing (at constant metabolism).

The formation of the so-called microenvironment around the body by clothing (textile layer/layers) was established. Thus, the increment of the thermal insulation of clothing I_{cl} made the skin temperature T_{sk} less and less dependent on the ambient temperature, due to the greater barrier between the body and the environment in terms of heat transfer.

It was found that the skin temperature was less dependent on the temperature asymmetry in the room as compared to the clothing temperature along the contour of the body. It was also found that with the increase in the activity to 3.0 met, the skin temperature became equal around the body contour and ceased to reflect the temperature asymmetry of the environment.

The existence of a nonlinear relationship between the augmentations of both skin and clothing temperature with the increase in the activity, while maintaining a constant clothing insulation, was established.

A functional relation between skin temperature and clothing temperature for different values of thermal insulation of clothing and/or metabolic activity was established. Such a relation was derived on the basis of a regression analysis of the data from the numerical simulation and did not appear originally in the Gagge's thermophysiological model.

An approach for deriving a set of regression equations was proposed, which could be enlarged even in the development of nomograms for the rapid determination of the temperature of the surface of the skin and clothing for various metabolic regimes and insulation of clothing in an environment with constant aerodynamic parameters.

References

AATCC 195. 2012. *Liquid Moisture Management Properties of Textile Fabrics*. American Association of Textile Chemists and Colorists, USA.

Abbott, N.J. 1973. The flammability of coated apparel fabrics. *J. Ind. Text.* 3: 135–41.

Adamson, A.W. 1990. *Physical Chemistry of Surfaces*, 5th ed., New York: John Wiley & Sons.

Akgun, M., B. Becerir, H. Alpay, S. Karaaslan, and A. Eke. 2010. Investigation of the effect of yarn locations on color properties of polyester automotive upholstery woven fabrics after abrasion. *Text. Res. J.* 80: 1422–31.

Al-Mogbel, A. 2003. A coupled model for predicting heat and mass transfer from a human body to its surroundings. In *Proceedings of the 36th AIAA Thermophysics Conference*, 724–32, Orlando, FL.

Alpay, H.R., B. Becerir, and M. Akgun. 2005. Assessing reflectance and color differences of cotton fabrics after abrasion. *Text. Res. J.* 75: 357–61.

Anderberg, A. and L. Wadsö. 2004. Moisture in self-levelling flooring compounds. Part II. Sorption isotherms. *Nordic Concr. Res.* 32: 16–30.

Angelova, R.A. 2003. Choosing the textiles: Indoor environmental criteria. In *Healthy Buildings*, 80–92, Sofia, Bulgaria: Avanguard.

Angelova, R.A. 2004. Thermal comfort and textile materials. In *Annual International Course* "Ventilation and Indoor Climate," 182–92. Sofia, Bulgaria: Avanguard.

Angelova, R.A. 2006. Textiles—An important factor for thermal comfort. In *Euro Academy on Ventilation and Indoor Climate* "Indoor Air Quality and Thermal Comfort," 210–19. Sofia, Bulgaria: Avanguard.

Angelova, R.A. 2007a. Maintaining the workers comfort and safety in extreme temperatures industrial environment. In *EuroAcademy on Ventilation and Indoor Climate* "Individually Controlled Environment," 197–205, Sofia, Bulgaria: Avanguard.

Angelova, R.A. 2007b. Phase change materials in intelligent textiles for individually maintained thermal environment. In *EuroAcademy on Ventilation and Indoor Climate* "Individually Controlled Environment," 73–8. Sofia, Bulgaria: Avanguard.

Angelova, R.A. 2008. The effects of textiles on thermal comfort. In *Proceedings of the Summer School on Build Environment*, 100–5, Sofia, Bulgaria: Avanguard.

Angelova, R.A. 2009. Moisture buffering materials in indoor environment. In *Proceedings of the 1st International Course* "Ventilation Efficiency and Indoor Climate Quality," 98–107, Ohrid, Macedonia.

Angelova, R.A. 2010a. Air-jet spinning of yarns. In *Advances in Yarn Spinning Technologies*, ed. C. Lawrence, Cambridge: Woodhead Publishing.

Angelova, R.A. 2010b. Problems of modelling of the air-permeability of textile fabrics with respect to their hierarchical structure. In *Proceedings of the 6th International Course* "Ventilation Efficiency and Indoor Climate Quality," 98–103, Pamporovo, Bulgaria: CERDECEN.

Angelova, R.A. 2010c. Pore approximation in numerical modeling of air-permeability of woven structures. In *Proceedings of the 2nd International Course on Ventilation Efficiency and Indoor Climate Quality*, pp. 115–20, Ohrid, Macedonia.

Angelova, R.A. 2010d. An image processing technique for measurement of the pore sizes in woven fabrics. In *Proceedings of the 2nd International Course on Ventilation Efficiency and Indoor Climate Quality*, pp. 109–14, Ohrid, Macedonia.

Angelova, R.A. 2011. Numerical procedure for prediction of the air-permeability of woven structures. In *Proceedings of the 7th International Course for Young Researchers*, Pamporovo, Bulgaria.

Angelova, R.A. 2012a. Air permeability of woven structures: Influence of the meso-structure. In *Proceedings of the 4th International Course "Ventilation Efficiency and Indoor Climate Quality,"* Ohrid, Macedonia: University of Skopije.

Angelova, R.A. 2012b. Determination of the pore size of woven structures through image analysis. *Cent. Eur. J. Eng.* 2: 129–35.

Angelova, R.A. 2013. Thermoregulation of a clothed human body: Physiology and thermophysiological models. In *Proceedings of the 5th International Course "Ventilation Efficiency and Indoor Climate Quality,"* 26–36, Ohrid, Macedonia: University of Skopije.

Angelova, R.A. and E. Nikolova. 2009. Experimental evaluation of the permeability of woven structures used for storage bags. In *Proceedings of the International Conference of the Faculty of Power Engineering, EMF 2009*, Sozopol, Bulgaria.

Angelova, R.A., G. Pichurov, I. Simova, P. Stankov, and I. Rodrigo. 2015. CFD based study of thermal sensation of occupants using thermophysiological model. Part II: Effect of metabolic rate and clothing insulation on human-environmental interaction. *Int. J. Cloth. Sci. Tech.* 27: 60–74.

Angelova, R.A., P. Stankov, I. Simova, and I. Aragon. 2011. Three dimensional simulation of air permeability of single layer woven structures. *Cent. Eur. J. Eng.* 1: 430–35.

Angelova, R.A., P. Stankov, I. Simova, and M. Kyosov. 2013. Computational modeling and experimental validation of the air permeability of woven structures on the basis of simulation of jet systems. *Text. Res. J.* 83: 1887–95.

Anon. 1994. *Woven & Knit Residential Upholstery Fabric Standard & Guidelines*, Joint Industry Fabric Standards Committee.

Anon. 2010. *Woven & Knit Residential Upholstery Fabric Standard & Guidelines*, Joint Industry Fabric Standards Committee.

Antunano, M.J. and S.A. Nunneley. 1992. Heat stress in protective clothing: Validation of a computer model and the heat-humidity index (HHI). *Aviat. Space Environ. Md.* 63: 1087–92.

Aoyagi, Y., T.M. McLellan, and R.J. Shephard. 1995. Determination of body heat storage in clothing. *Eur. J. Appl. Physiol.* 71: 197–206.

Armour, J. and J. Cannon. 1968. Fluid flow through woven screens. *AIChE J.* 14: 415–20.

Ashpole, D.K. 1952. The moisture relations of textile fibres at high humidities. *Proc. R. Soc. Lond. A Math. Phys. Sci.* 212(1108): 112–23.

ASHRAE. 1992. *Thermal Environmental Conditions for Human Occupancy*. ANSI/ASHRAE Standard 55–1992.

ASHRAE. 1993. *Handbook of Fundamentals*. American Society of Heating, Refrigeration and Air-Conditioning Engineers, New York: American Society of Heating, Refrigeration and Air-Conditioning Engineers.

ASHRAE. 1997. *Handbook of Fundamentals: Physiological Principles, Comfort, Health*. American Society of Heating, Refrigeration and Air-Conditioning Engineers, New York.

ASHRAE. 2005. *Handbook of Fundamentals*. New York: American Society of Heating, Refrigeration and Air-Conditioning Engineers.

ASTM E1530-04. 2004. *Standard Test Method for Evaluating the Resistance to Thermal Transmission of Materials by the Guarded Heat Flow Meter Technique*. American Society for Testing and Materials, USA.

Backer, S. 1951. The relationship between the structural geometry of a textile fabric and its physical properties. Part IV: Interstice geometry and air permeability. *Text. Res. J.* 21: 703–14.

Baitinger, W.F. 1979. Product engineering of safety apparel fabrics: Insulation characteristics of fire-retardant cottons. *Text. Res. J.* 49: 221–5.

Baltakytė, R. and S. Petrulytė. 2008. Experimental analysis of air permeability of terry fabrics with hemp and linen pile. *Mater. Sci.* (MEDŽIAGOTYRA) 14: 258–62.

Barker, R.L. 2002. From fabric hand to thermal comfort: The evolving role of objective measurements in explaining human comfort response to textiles. *Int. J. Cloth. Sci. Tech.* 14: 181–200.

Bartels, V.T. 2003. Thermal comfort of aeroplane seats: Influence of different seat materials and the use of laboratory test methods. *Appl. Ergon.* 34: 393–99.

Barnes, J. and B. Holcombe. 1996. Moister sorption and transport in clothing during wear. *Text. Res. J.* 66: 703–714.

Basal, G. and W. Oxenham. 2006. Comparison of properties and structures of compact and conventional spun yarns. *Text. Res. J.* 76: 567–75.

Behera, B.K. and P.K. Hari. 2010. *Woven Textile Structure*. New York: CRC Press.

Belding, H.S. and T.F. Hatch. 1956. Index for evaluating heat stress in terms of resulting physiological strain. *ASHRAE Tran.* 62: 213–36.

Belov, E.B., S.V. Lomov, I. Verpoest, T. Peters, D. Roose, R.S. Parnas, K. Hoes, and H. Sol. 2004. Modelling of permeability of textile reinforcements: Lattice–Boltzmann method. *Compos. Sci. Technol.* 64: 1069–80.

Benisek, L., G.K. Edmondson, and W.A. Phillips. 1979. Protective clothing–Evaluation of wool and other fabrics. *Text. Res. J.* 49: 212–21.

Benzinger, T.H. 1969. Heat regulation: Homeostasis of central temperature in man. *Physiol. Rev.* 49: 671–759.

Berger-Preiss, E., K. Levsen, G. Leng, H. Idel, D. Sugiri, and U. Ranft. 2002. Indoor pyrethroid exposure in homes with woolen textile floor coverings. *Int. J. Hyg. Environ. Health* 205: 459–72, 2002.

Bhattacharjee, D. and V.K. Kothari. 2008. Prediction of thermal resistance of woven fabrics. Part II: Heat transfer in natural and forced convective environments. *J. Text. I.* 99: 433–49.

Bilisik, K., Y. Turhan, and O. Demiryurek. 2011. Tearing properties of upholstery flocked fabrics. *Text. Res. J.* 81: 290–300.

Bilisik, K. and G. Yolacan. 2009. Abrasion properties of upholstery flocked fabrics. *Text. Res. J.* 79: 1625–32.

Boestra, A.C. and J.L. Leyton. 1997. Diagnosing problem buildings: The risk factor approach. *Indoor Air* 2: 278–83.

Bornehag, C.G., J. Sundell, and L. Hägerhed. 2003. Dampness in dwellings and sick building symptoms among adults: A cross-sectional study on 8918 Swedish homes. In *Proceedings of the Healthy Buildings*, Singapore: National University of Singapore, Vol. 1, 582–87.

Brebner, D.F., D. McKerslake, and J.L. Wadell. 1956. The diffusion of water vapor through human skin. *J. Physiol.* 132: 225–31.

Bruschke, M. and S.G. Advani. 1993. Flow of generalized Newtonian fluids across periodic array of cylinders. *Rheology* 37: 479–97.

Bryant, Y.G. and D.P. Colvin. 1992. Fibers with enhanced, reversible thermal energy storage properties. In *Techtextil Symposium*, Frankfurt, Germany, 1–8.

Bullard, R.W., M.R. Baneijee, F. Chen, R. Elizondo, and B.A. MacIntyre. 1970. Skin temperature and thermoregulatory sweating: A control system approach. In *Physiological and Behavioral Temperature Regulation*, eds. J.D. Hardy, A.P. Gagge, and J.A.J. Stolwijk. Springfield, IL: Charles C Thomas, 597–610.

Burton, A.C. 1941. The operating characteristics of the human thermoregulatory mechanism. In *Temperature: Its Measurement and Control in Science and Industry*, New York: Reinholds Publishing Corporation.

Burton, A.E. and O.G. Edholm. 1955. *Man in a Cold Environment*, London: Edward Arnold Publishing.

Butte, W. 2003. Reference values of environmental pollutants in house dust. In *Indoor Environment: Airborne Particles and Settled Dust*, eds. L. Morawska and T. Salthammer, Weinheim, Germany: Wiley-VCH Verlag GmbH, pp. 407–435.

Cai, Z. and A.L. Berdichevsky. 1993. An improved self-consistent method for estimating the permeability of a fiber assembly. *Polym. Compos.* 14: 314–23.

Carman, P. 1956. *Flow of Gases thorugh Porous Media*, New Year: Academic Press.

Carmeliet, J., M. de Wit, and H. Janssen. 2005. Hysteresis and moisture buffering of wood. In *Symposium of Building Physics in the Nordic Countries*, 55–62, Reykjavik, Iceland.

Carvalho, P.M., M. Gameiro da Silva, and J. Ramos. 2013. Influence of weather and indoor climate on clothing of occupants in naturally ventilated school buildings. *Build. Environ.* 59: 38–46.

Catalli, V. 1995. Choosing carpets for good IAQ. *Proc. Peace Env.* 6: 235–9.

Cay, A., S. Vassiliadis, M. Rangoussi, and I. Tarakcioglu. 2004. On the use of image processing techniques for the estimation of the porosity of textile fabrics. *Int. J. Signal Proc.* 1: 51–54.

CEN/TR 16298. 2011. Textiles and textile products—Smart textiles—Definitions, categorisation, applications and standardization needs. Belgium: European Committee for Standardization.

Cheng, Y., J. Niu, and N. Gao. 2012. Thermal comfort models: A review and numerical investigation. *Build. Environ.* 47: 13–22.

Chow, T.T., A. Kwan, Z. Lin, and W. Bai. 2006. Conversion of operating theatre from positive to negative pressure environment. *J. Hosp. Infect.* 64: 371–78.

Clayton, F.H. 1935. The measurement of the air permeability of fabrics. *J. Text. I Trans.* 26: T171–86.

Collier, B.J. and H.H. Epps. 1999. *Textile Testing and Analysis*, Upper Saddle River, NJ: Prentice Hall.

Corbelini, E. 1987. Comfort in clothing for outdoor activities and multifilament polypropylene yarn. *Chemiefasern/Textilindustrie* 37/89. T39-T47.

Cui, P. and F. Wang. 2009. An investigation of heat flow through kapok insulating material. *Tekst Konfeksiyon* 2: 88–92.

D'Ambrosio, F.R. 2006. Human body thermoregulation models. In *EuroAcademy on Ventilation and Indoor Climate* "Indoor Air Quality and Thermal Comfort," 210–9. Sofia, Bulgaria: Avanguard Publishing.

Daukantienė, V. and A. Skarulskienė. 2005. Wear and hygienic properties of cotton velvet fabrics. In *Proceedings of the Radom International Scientific Conference Radom, Poland*, Poland: Technical University of Radom, 355–58.

David, H.G. and P. Nordon. 1939. Case studies of coupled heat and moisture diffusion in wool beds. *Text. Res. J.* 39: 166–72.

Davies, C.N. 1952. The separation of airborne dust and particles. *P. I. Mech. Eng.* 18: 185–213.

Debnath, S. and M. Madhusoothanan. 2011. Thermal resistance and air permeability of jute-polypropylene blended needle-punched nonwoven. *Indian J. Fibre Text.* 36: 122–31.

Degen, K.G., S. Rosetto, T. Reinchenauer, and J. Fricke. 1992. Investigation of the heat transfer in evacuated foil spacer multilayer insulation. *J. Build. Phys.* 16: 140–52.

Delerue, J.-F., S. Lomov, V. Parnas, R.S. Verpoest, and I. Wevers. 2003. Pore network modeling of permeability for textile reinforcements. *Polym. Compos.* 24: 344–57.

Dhingra, R.C. and R. Postle. 1977. Air permeability of woven, double-knitted outer-wear fabrics. *Text. Res. J.* 47: 630–31.

Ding D., T. Tang, G. Song, and A. McDonald. 2010. Characterizing the performance of a single-layer fabric system through a heat and mass transfer model–Part I: Heat and mass transfer model. *Text. Res. J.* 81: 398–411.

Ding D., T. Tang, G. Song, and A. McDonald. 2011. Characterizing the performance of a single-layer fabric system through a heat and mass transfer model–Part II: Thermal and evaporative resistances. *Text. Res. J.* 81: 945–58.

Djonov, E. and T. Van Leeuwen. 2011. The semiotics of texture: From tactile to visual. *Visual Commun.* 10: 541–64.

Donahaye, E.J., S. Navarro, C. Bell, D. Jayas, R. Noyes, and T.W. Phillips (eds.). 2007. An evaluation of the permeability to phosphine through different polymers used for the bag storage of grain. In *Proceedings of the International Conference on Controlled Atmosphere and Fumigation in Stored Products*, Gold Coast, Australia.

Donaldson, D.J., H. Mard, and R.J. Harper, Jr. 1979. Imparting smolder-resistance to cotton upholstery fabric. *Text. Res. J.* 49: 185–90.

Donaldson, D.J., D. Yeadon, and R.J. Harper, Jr. 1981. Smoldering characteristics of cotton upholstery fabrics. *Text. Res. J.* 51: 196–202.

Douglas, R. and M. Huiping. 1992. On the derivation of the Forchheimer equation by means of the averaging theorem. *Transport Porous Med.* 7: 255–64.

DuBois, D. and E.F. DuBois. 1916. A formula to estimate the approximate surface area if height and weight be known. *Arch. Intern. Med.* 17: 863–71.

Dubrovski, P.D. 2000. Volume porosity of woven fabrics. *Text. Res. J.* 70: 915–19.

Dubrovski, P.D. 2001. A geometrical method to predict the macroporosity of woven fabrics. *J. Text. I.* 92: 288–98.

Dubrovski, P.D. and M. Brezocnik. 2002. Using genetic programming to predict the macroporosity of woven cotton fabrics. *Text. Res. J.* 72: 187–94.

Dubrovski, P.D. and M. Sujica. 1995. The connection between woven fabric construction parameters and air permeability. *Fibers Text. East. Eur.* 3: 37–41.

Duru Cimilli, S., E. Deniz, C. Candan, and B.U. Nergis. 2012. Determination of natural convective heat transfer coefficient for plain knitted fabric via CFD modeling. *Fibers Text. East. Eur.* 20: 42–6.

Elnashar, E.A. 2005. Volume porosity and permeability in double-layer woven fabric. *AUTEX Res. J.* 5: 207–17.

EN 13795. 2006. Surgical Drapes, Gowns and Clean Air Suits, Used as Medical Devices for Patients, Clinical Staff and Equipment – General Requirements for Manufacturers, Processors and Products, Test Methods, Performance Requirements and Performance Levels. Switzerland: International Organization for Standardization.

EN ISO 1973. 1999. *Textile Fibres-Determination of Linear Density—Gravimetric Method and Vibroscope Method*. Switzerland: International Organization for Standardization.

EN ISO 2061. 2010. *Textiles: Determination of Twist in Yarns—Direct Counting Method*.

EN ISO 2062. 2010. *Textiles: Yarns from Packages—Determination of Single-End Breaking Force and Elongation at Break Using Constant Rate of Extension (CRE) Tester.* Switzerland: International Organization for Standardization.

EN ISO 5084. 2002. *Textiles: Determination of Thickness of Textiles and Textile Products.* Switzerland: International Organization for Standardization.

EN ISO 9237. 1999. *Textiles: Determination of Permeability of Fabrics to Air.* Switzerland: International Organization for Standardization.

Epps, H. 1988. Insulation characteristics of fabric assemblies. *J. Coated Fabrics* 17: 212–9.

Epps, H.H. 1996. Prediction of single-layer fabric air permeability by statistics modeling. *J. Test Eval.* 24: 26–31.

Epps, H.H. and K.K. Leonas. 1997. The relationship between porosity and air permeability of woven textile fabrics. *J. Test. Eval.* 25: 108–13.

Epps, H.H. and M.K. Song. 1992. Thermal transmittance and air permeability of plain weave fabrics. *Cloth. Text. Res. J.* 11: 10–17.

Epstein, Y., Y. Heled, I. Ketko, J. Muginshtein, R. Yanovich, A. Druyan, and D.S. Moran. 2013. The effect of air permeability characteristics of protective garments on the induced physiological strain under exercise-heat stress. *Ann. Occup. Hyg.* 57: 866–74.

European Medical Devices Directive 2007/47/EC. 2007. DIRECTIVE 2007/47/EC of the European Parliament and of the Council. *Off. J. Eur. Union* 21.09, Vol. 50.

Fabbri, K. 2013. Thermal comfort evaluation in kindergarten: PMV and PPD measurement through datalogger and questionnaire. *Build. Environ.* 68: 202–14.

Fan, J. and J. He. 2012. Fractal derivative model for air permeability in hierarchic porous media. *Abstract and Applied Analysis*, Article ID 354701, 7 pages.

Fang, L., G. Clausen, and P.O. Fanger. 2000. Temperature and humidity: Important factors for perception of air quality and for ventilation requirements. *ASHRAE Trans.* 106: 503–10.

Fanger, P.O. 1967. Calculation of thermal comfort: Introduction of a basic comfort equation. *ASHRAE Tran.* 73(2): 1–20.

Fanger, P.O. 1972. *Thermal Comfort,* New York: McGraw-Hill Book Company.

Faris, B.F. 1995. *Automotive Textiles,* ed. M. Ravnitzky, SAE, Warrendale, PA.

Farnworth, B. 1983. Mechanisms of heat flow through clothing insulation. *Text. Res. J.* 53: 717–25.

Farnworth, B. 1986. Numerical model of the combined diffusion of heat and water vapour through clothing. *Text. Res. J.* 56: 653–65.

Fatahi, II. and A. Alamdar. 2010. Assessment of the relationship between air permeability of woven fabrics and its mechanical properties. *Fibers Text East Europe* 18: 68–71.

Fiala, D., K.J. Lomas, and M. Stohrer. 1999. A computer model of human thermoregulation for a wide range of environmental conditions: The passive system. *J. Appl. Physiol.* 87(5): 1957–72.

Fiala, D., A. Psikuta, G. Jendritzky, S. Paulke, D.A. Nelson, W. Lichtenbelt, and A. Frijns. 2010. Physiological modeling for technical, clinical and research applications. *Front Biosci.* 2: 939–68.

Fleural-Lessard, F. and B. Serrano. 1990. Resistance to insect perforation of plastic films for stored—Product packing: Methodological study on tests with rice weevil and layer grain borer. *Sci. Aliments* 10: 521–32.

Fobelets, A.P.R. and A.P. Gagge. 1988. Rationalization of the effective temperature ET*, as a measure of the enthalpy of the human indoor environment. *ASHRAE Trans.* 94(1): 12–31.

Fohr, J.P., D. Couton, and G. Treguier. 2002. Dynamic heat and water transfer through layered fabrics. *Text. Res. J.* 72: 1–12.

Foulger, S.H. and R.V. Gregory (Clemson). 2003. Intelligent Textiles Based on Environmentally Responsive in Fibers. National Textile Center Annual Report, South Carolina: Clemson University.

Fourt, L.E. and N.R.S. Hollies. 1970. *Clothing and Comfort*, New York: Marcel Dekker.

Fricke, J. and R. Caps. 1988. Heat transfer in thermal insulations—Recent progress and analysis. *Int. J. Thermophys.* 9: 885–93.

Frydrych, I., G. Dziworska, and J. Bilska. 2002. Comparative analysis of the thermal insulation properties of fabrics made of natural and man-made cellulose fibres. *Fibers Text East Europe* 10: 40–45.

Fung, W. 2010. Textiles in transportation. In *Handbook of Technical Textiles*, eds., A.R. Horrocks and S.C. Anand, Cambridge: Woodhead Publishing.

Fung, W. and M. Hardcastle. 2001. *Textiles in Automotive Engineering*, Cambridge: Woodhead Publishing.

Gagge, A.P. 1937. A new physiological variable associated with sensible and insensible perspiration. *Am. J. Physiol.* 120: 277–87.

Gagge, A.P. 1971. A two-node model of human temperature regulation. In *FORTRAN Bioastronautics Data Book*, 142–8, eds. J.F. Parker, Jr. and V.R. West, Washington, DC: NASA SP-3006.

Gagge, A.P. 1973. A two-node model of human temperature regulation in FORTRAN, In *Bioastronautics Data Book*, eds. J.F. Parker, Jr. and V.R. West, NASA SP-3006, Washington, DC, 142–48.

Gagge, A.P., A.C. Burton, and H.C. Bazett. 1941. A practical system of units for the description of the heat exchange of man with his environment. *Science* 94: 428–30.

Gagge, A.P., A.P. Fobelets, and L.G. Berglund. 1986. A standard predictive index of human response to the thermal environment. *ASHRAE Trans.* 92: 709–31.

Gagge, A.P., R. Richard, and M. Gonzalez. 1996. Mechanisms of heat exchange: Biophysics and physiology. In *Comprehensive Physiology 2011, Supplement 14: Handbook of Physiology, Environmental Physiology*, New York: Oxford University Press, 45–84.

Gagge, A.P., J.A. Stolwijk, and J.D. Hardy. 1967. Comfort and thermal sensations and associated physiological responses at various ambient temperatures. *Environ. Res.* 1: 1–20.

Gagge, A.P., J.A.J. Stolwijk, and Y. Nishi. 1969. The prediction of thermal comfort when thermal equilibrium is maintained by sweating. *ASHRAE Trans.* 75: 108–25.

Gagge, A.P., J.A.J. Stolwijk, and Y. Nishi. 1971. An effective temperature scale, based on a simple model of human physiological regulatory response. *ASHRAE Trans.* 77: 247–62.

Gagge, A.P., C.E.A. Winslow, and L.P. Herrington. 1938. The influence of clothing on physiological reactions of the human body to varying environmental temperatures. *Am. J. Physiol.* 124: 30–50.

Gandhi, S. and S.M. Spivak. 1994. A survey of upholstered furniture fabrics and implications for furniture flammability. *J. Fire Sci.* 12: 284–312.

Gao, N. and J. Niu. 2004. CFD study on micro-environment around human body and personalized ventilation. *Build. Environ.* 39: 795–805.

Gebart, B.R. 1992. Permeability of unidirectional reinforcements for RTM. *J. Compos. Mater.* 26: 1100–33.

Gebhart, B. 1993. *Heat Conduction and Mass Diffusion.* New York: McGraw-Hill.

Ghali, K., B. Jones, and J. Tracy. 1995. Modelling heat and mass transfer in fabrics. *Int. J. Heat Mass Tran.* 38: 13–21.

Ghali, K., N. Ghaddar, and B. Jones. 2002. Empirical evaluation of convective heat and moisture transport coefficients in porous cotton medium. *J. Heat. Transf.* 124: 530–37.

Gibson, Ph., D. Rivin, C. Kendrick, and H. Gibson. 1999. Humidity dependant air permeability of textile materials. *Text. Res. J.* 69: 311–17.

Goodfellow, H.D. 2001. *Industrial Ventilation Design Guidebook,* London: Academic Press.

Goodings, A.G. 1964. Air flow through textile fabrics. *Text. Res. J.* 34: 713–24.

Gooijer, H., M.M.C.G. Warmoeskerken, and J. Groot Wassink. 2003a. Flow resistance of textile materials. Part I: Monofilament fabrics. *Text. Res. J.* 73: 437–43.

Gooijer, H., M.M.C.G. Warmoeskerken, and J. Groot Wassink. 2003b. Flow resistance of textile materials. Part II: Multifilament fabrics. *Text. Res. J.* 73: 480–84.

Gravesen, S., P. Skov, O. Valbørn, and H. Lowenstein 1990. The role of potential immunogenic components of dust (mod) in the sick-building syndrome. *Indoor Air* 1: 9–13.

Grouve, W.J.B., R. Akkerman, R. Loendersloot, and S. van den Berg. 2008. Transverse permeability of woven fabrics. In *Proceedings of the 11th ESAFORM Conference on Material Forming,* Lyon, France.

Guillaume, E., C. Chivas, and A. Sainrat. 2012. Regulatory issues and flame retardant usage in upholstery furniture in Europe. In *Fire and Building Safety in the Single European Market,* Edinburgh: School of Engineering and Electronics, University of Edinburgh, 38–48.

Guo, H. and F. Murray. 2000. Modelling of emissions of total volatile organic compounds in an australian house. *Indoor Built. Environ.* 9: 171–81.

Guo, Z. and C. Shu. 2013. *Advances in Computational Fluid Dynamics: Vol. 3, Lattice Boltzmann Method and Its Applications in Engineering,* Singapore: World Scientific Publishing.

Gutowski, T.G., Z. Cai, S. Bauer, D. Boucher, J. Kingery, and S. Wineman. 1987. Consolidation experiments for laminate composites. *J. Compos. Mater.* 21: 650–69.

Habboub, A.K. 2003. Thermal evaluation of body support systems using thermogrammetry and interfacial temperature sensing. In *Proceedings of the International Conference on Thermal Engineering and Thermogrammetry,* Budapest, Hungary.

Haghi, A.K. 2002. Mechanism of heat and mass transfer in moist porous materials. *J. Teknologi.* 36(F): 1–14.

Hamilton, J.B. 1964. A general system of woven-fabric geometry. *J. Text. I.* 55: 66–82.

Hancock, P.A., J.M. Ross, and J.L. Szalma. 2007. A meta-analysis of performance response under thermal stressors. *J. Hum. Fac. Erg. Soc.* 49: 851–77.

Harderup, L.-E. 1999. Hygroscopic moisture of the indoor air considering non-stationary phenomena. In *Synthesis of Publications for the Period 1979–1998,* Lund, Sweden.

Harderup, L.-E. 2005. A PC-model to predict moisture buffer capacity in building materials according to a NORDTEST method. In *Nordic Building Physics Symposium,* Reykjavik, Iceland.

Hardy, J.D. 1972. Models of temperature regulation—A review. In *Essays on Temperature Regulation,* Amsterdam, the Netherlands: North Holland Publishing.

Hardy, J.D. and J.A.J. Stolwijk. 1966. Partitional calorimetry studies of man during exposure to thermal transients. *J. Appl. Physiol.* 21: 1799–806.

Harper, R.J., Jr., G. Ruppenicker, Jr., and D. Donaldson. 1986. Cotton blend fabrics from polyester core yarns. *Text. Res. J.* 56: 80–6.

Hatch, K.L. 1993. *Textile Science.* Minneapolis, MN: West Publishing.

Havenith, G. 2001. Individualized model of human thermoregulation for the simulation of heat stress response. *J. Appl. Physiol.* 90: 1943–54.

Havenith, G. 2002. Interaction of clothing and thermoregulation. *Exogenous Dermatol.* 1: 221–30.

Havenith, G. 2003. Textiles and the skin. In *Current Problems in Dermatology*, ed. P. Elsner et al., Basel, Switzerland: Karger AG.

Havlova, M. 2010. Influence of vertical porosity on woven fabric air permeability. In *7th International Conference—TEXSCI*, Liberec, Czech Republic.

Henry, P.S.H. 1939. Diffusion in absorbing media. *Proc. R. Soc.* 171A: 215–41.

Henry, P.S.H. 1948. Diffusion of moisture and heat through textiles. *Discuss. Faraday Soc.* 3: 243–57.

Hensel, H. 1981. *Thermoreception and Temperature Regulation*, London: Academic Press.

Hettich, B.V. 1984. Performance characteristics and applications of flame resistant rayon fabrics. *Text. Res. J.* 54: 382–90.

Highland, H.A. 1981. Resistant barriers for stored- product insects. In *Handbook of Transportation and Marketing in Agriculture, Vol. 1: Food Commodities*, ed. E. Finney, New York: CRC Press, 41–45.

Hohenstein Test Report. 2000. *Fabrics for Surgical Gowns and OR-Garment Ensembles*, Z0.4.3884, Germany: Bekleidungs physiologisches Institut Hohenstein.

Holcombe, B.V. and B.N. Hoschke. 1983. Dry heat transfer characteristics of underwear fabrics. *Text. Res. J.* 53: 368–74

Holland, E.J., C.A. Wilson, and B.E. Niven. 1999. Microclimate ventilation of infant bedding. *Int. J. Cloth. Sci. Tech.* 11: 226–39.

Holopainen, R. 2012. *A Human Thermal Model for Improved Thermal Comfort*, PhD thesis, Aalto University, Finland.

Hossain, M., M. Acar, and W. Malalasekera. 2005. A mathematical model for airflow and heat transfer through fibrous webs. *J. Process. Mech. Eng. E.* 219: 357–66.

Houdas, Y. 1981. Modelling of heat transfer in man. In *Bioengineering, Thermal Physiology and Comfort*, Amsterdam, the Netherlands: Elsevier Scientific Publishing.

Houghton, F.C. and C.P. Yaglou. 1923. Determining lines of equal comfort. *ASH&VE Trans.* 28: 163–76.

Howorth, W.S. and P.H. Oliver. 1958. The application of multiple factor analysis to the assessment of fabric handle. *J. Text. I.* 49: T540–53.

Hsieh, Y.L. 1995. Liquid transport in fabric structures. *Text. Res. J.* 65: 299–307.

Huizenga, C., Z. Hui, and E. Arens. 2001. A model of human physiology and comfort for assessing complex thermal environments. *Build Environ.* 36: 691–9.

Ihrig, A.M., A.L. Rhyne, V. Norman, and A.W. Spears. 1986. Factors involved in the ignition of cellulosic upholstery fabrics by cigarettes. *J. Fire Sci.* 4: 237–60.

Ilce, A.C. and B. Cayir. 2010. Determination of bus passenger environmental ergonomics expectations and preferences. *Technology* 13: 261–71.

Ilmarinen, R. 2005. Combat clothing—A thermal barrier. In *Proceedings of the International Congress on Soldiers' Physical Performance*, Jyväskylä, Finland.

Ishtiaque, S.M., A. Das, and A.K. Kundu. 2014. Clothing comfort and yarn packing relationship: Part II—Transmission characteristics of fabrics. *J. Text. I.* 105: 736–43.

ISO 3801. 1977. *Textiles: Woven Fabrics—Determination of Mass per Unit Length and Mass per Unit Area.* Switzerland: International Organization for Standardization.

ISO 7211-2. 1984. *Textiles: Woven Fabrics—Construction—Methods of Analysis. Part 2: Determination of Number of Threads per Unit Length.* Switzerland: International Organization for Standardization.

ISO 7211-3. 1984. *Textiles: Woven Fabrics—Construction—Methods of Analysis. Part 3: Determination of Crimp of Yarn in Fabric.* Switzerland: International Organization for Standardization.

ISO 7730. 1995. International Standard, Moderate Thermal Environments—Determination of the PMV and PPD Indices and Specification of the Conditions for Thermal Comfort. Switzerland: International Organization for Standardization.

ISO 9920. 1995. International Standard: Ergonomics of the Thermal Environment—Estimation of the Thermal Insulation and Evaporative Resistance of Clothing Ensembles, Switzerland: International Organization for Standardization.

Ivanov, M., P. Stankov, D. Markov, I. Simova, N. Kehaiov, and E. Georgiev. 2014. A study on performance assessment of students in controlled indoor environment conditions. In *Proceedings of the International Conference Sustainable Solutions for Energy and Environment*, Bucharest, Romania.

IWTO (International Wool Textile Organisation). 2010. *Wool for Interior Textiles, Brochure.* www.iwto.org/files/publications/19.pdf.

Jakšić, D. 1975. The development of the new method to determine the pore size and pore size distribution in textile products. Part I. In *Proceedings of the Faculty of Natural Sciences and Technology*, 215–23, Ljubljana, Slovenia.

Jakšic, D. and N. Jakšic. 2007. Assessment of porosity of flat textile fabrics. *Text. Res. J.* 77: 105–10.

Jeremy, D., D. Reversat, and A. Gagalowicz. 2003. Modelling hysteretic behaviour of fabrics. In *Proceedings of the Computer Vision/Computer Graphics Collaboration for Model-based Imaging*, Rocquencourt (France).

Jerkovic, I., J.M. Pallares, and X. Capdevila. 2010. Study of the abrassion resistance in the upholstery of automobile seats. *AUTEX Res. J.* 10: 14–20.

Johnson, R.W. 1998. *The Handbook of Fluid Dynamics*, New York: CRC Press.

Joint Industry Fabric Standards Committee. 2010. *Woven & Knit Residential Upholstery Fabric Standard & Guidelines.*

Jones, B.W. 2002. Capabilities and limitations of thermal models for use in thermal comfort standard. *Energ. Buildings.* 34: 653–59.

Kajiwara, N., J. Desborough, S. Harrad, and H. Takigami. 2013. Photolysis of brominated flame retardants in textiles exposed to natural sunlight. *Environ. Sci. Proc. Impacts* 15: 653–60.

Kemp, A. 1958. An extension of Pierce's cloth geometry to the treatment of non-circular threads. *J. Text. I.* 49: 44–9.

Kennedy, M. 2002. Carpeting/Flooring. *Ame. Sch. Uni.* 75(4): 44–7.

Kidesø, J., P. Vinzents, and T. Schneider. 1999. A simple method for measuring the potential resuspension of dust from carpets in the indoor environment. *Text. Res. J.* 69: 169–75.

Kolodyazhniy, V., J. Späti, S. Frey, T. Götz, A. Wirz-Justice, K. Kräuchi, C. Cajochen, and F.H. Wilhelm. 2011. Estimation of human circadian phase via a multi-channel ambulatory monitoring system and a multiple regression model. *J. Biol. Rhythm.* 26: 55–67.

Konova, H. and R.A. Angelova. 2013. False-Twister as a yarn formation factor in wrap spinning, *Text. Res. J.* 83: 1926–35.

Kosaka, M., M. Yamane, R. Ogai, T. Kato, N. Ohnishi, and E. Simon. 2004. Human body temperature regulation in extremely stressful environment. *J. Therm. Biol.* 29: 495–501.

Kothari, V.K. and D. Bhattacharjee. 2008. Prediction of thermal resistance of woven fabrics. Part I: Mathematical model. *J. Text. I* 99: 421–32.

Kothari, V.K. and A. Newton. 1974. The air-permeability of non-woven fabrics. *J. Text. I.* 65: 525–31.

Kotresh, T.M. 1996. Hazard potential of apparel textiles. *Indian J. Fibre Text.* 21: 157–60.

Kozeny, J. 1927. *Uber Kapillare heitung des Wassers in Boder.* Vienna, Austria: Royal Academy of Science.

Kulichenko, A.V. 2005. Theoretical analysis, calculation, and prediction of the air permeability of textiles. *J. Fibre. Chem.* 37: 371–80.

Kulichenko, A. and L. Van Langenhove. 1992. The resistance to flow transmission of porous materials. *J. Text. I* 83: 127–32.

Kullman, R.M.H., C.O. Graham, and G.F. Ruppenicker. 1981. Air permeability of fabrics made from unique and conventional yarns. *Text. Res. J.* 51: 781–86.

Kyeyoun, Ch., G. Cho, P. Kim, and Ch. Cho. 2004. Thermal storage/release and mechanical properties of phase change materials on polyester fabrics. *Text. Res. J.* 74: 292–96.

Lain, R.C. and J.L. Kardos. 1988. The permeability of aligned and cross-plied fiber beds during processing of continuous fiber composites. In *Proceedings of the American Society for Composites, Third Technical Conference*, 3–11, Washington, DC: University of Dayton.

Lamb, G.E.R. and P.A. Costanza. 1979. Influences of fiber geometry on the performance of nonwoven air filters. *Text. Res. J.* 49: 79–87.

Le, C.V. and N.G. Ly. 1995. Heat and mass transfer in the condensing flow of stream through an absorbing fibrous medium. *Int. J. Heat. Mass. Trans.* 38: 81–9.

Lee, H.S., W.W. Carr, H.W. Beckham, and W.J. Weper. 2000. Factors influencing air flow through unbacked tufted carpets. *Text. Res. J.* 70: 876–85.

Levin, H. 1989. Building materials and indoor air quality. In *State of the Art Reviews in Occupational Medicine*, eds. M. Hodgson and J. Cone, Philadelphia, PA: Hanley and Belfus.

Lewis, R.G., R.C. Fortmann, and D.E. Camann. 1994. Evaluation of methods for monitoring the potential exposure of small children to pesticides in the residential environment. *Arch. Environ. Con. Tox.* 26.1: 37–46.

Li, Y. 2007. Computational textile bioengineering. *Stud. Comp. Intell.* 55: 203–21.

Li, Y. and B.V. Holcombe. 1992. A two-stage sorption model of the coupled diffusion of moisture and heat in wool fabrics. *Text. Res. J.* 62: 211–7.

Li, Y. and B.V. Holcombe. 1998. Mathematical simulation of heat and moisture transfer in a human-clothing-environment system. *Text. Res. J.* 68: 389–97.

Li, J., Zh. Zhang, and Y. Wang. 2013. The relationship between air gap sizes and clothing heat transfer performance. *J. Text. I.* 104: 1327–36.

Lis, S.J., G. Engholm, and R.A. Bambenek. 1962. Thermal conductivity of nylon parachute fabrics subjected to compressional stresses. *Text. Res. J.* 32: 24–8.

Licina, D., J. Pantelic, A. Melikov, Ch. Sekhar, and K.W. Tham. 2014. Experimental investigation of the human convective boundary layer in a quiescent indoor environment. *Build. Environ.* 75: 79–91.

Lind, A.R. 1970. Effect of individual variation on upper limit of prescriptive zone of climates. *J. Appl. Physiol.* 28: 57–62.

Lomov, S.V., D. Ivanov, and I. Verpoest. 2010. Predictive models for textile composites. In *Proceedings of the 7th International Conference—TEXSCI 2010*, Liberec, Czech Republic.

Lord, J. 1959. Determination of the air permeability of fabrics. *J. Text. I* 50: T569–82.

Lotens, W.A. 1988. Comparison of thermal predictive models for clothed humans. *ASHRAE Tran.* 94(1): 1321–40.

Lotens, W.A. and G. Havenith. 1991. Calculation of clothing insulation and vapor resistance. *Ergonomics* 34: 233–54.

Love, L. 1954. Graphical relationships in cloth geometry for plain, twill and sateen weaves. *Text. Res. J.* 24: 1073–83.

Lozanova, M., P. Stankov, and J. Denev. 1997. Turbulent kinetic energy and its dissipation in rectangular jets—Experimental assessment with respect to CFD modelling. In *Conference on Finite Difference Methods: Theory and Applications*, Rousse, Bulgaria, 167–73.

Lozanova, M. and P. Stankov. 1998. Experimental investigation on the similarity of a 3D rectangular turbulent jet. *Exp. Fluids* 24: 470–78.

Luo, Y., I. Verpoest, K. Hoes, M. Vanheule, H. Sol, and A. Cardon. 2001. Permeability measurement of textile reinforcements with several test fluids. *Composites Part A*. 32: 1497–504.

Mao, N. 2009. Permeability in engineering non-wovens fabrics having patterned structure. *Text. Res. J.* 79: 1348–57.

Markov, D., P. Stankov, M. Ivanov, N. Kehaiova, and E. Georgiev. 2014. On the influence of indoor temperature on occupant's performance. In *Proceedings of the International Scientific Conference on Ruse*, Ruse, Bulgaria.

McCullough, E.A., B.W. Jones, and J. Huck. 1985. A comprehensive data base for estimating clothing insulation. *ASHRAE Trans.* 91: 29–47.

McCullough, E.A., B.W. Jones, and T. Tamura. 1989. A data base for determining the evaporative resistance of clothing. *ASHRAE Trans.* 95(II): 316–28.

McCullough, E.A., B.W. Olesen, and S. Hong. 1994. Thermal insulation provided by chairs. *ASHRAE Tran.* 100: 795–802.

Melikov, A. 2006. Thermal comfort prediction: Conditions for thermal comfort. In *EuroAcademy on Ventilation and Indoor Climate* "Indoor Air Quality and Thermal Comfort," 13–26, Sofia, Bulgaria: Avanguard.

Militky, J., M. Vik, M. Vikova, and D. Kremenakova. 2010. *Influence of Fabric Construction on the Their Porosity and Air Permeability*, http://centrum.tul.cz/centrum/centrum/1Projektovani/1.2_publikace/ %5B1.2.30%5D.pdf.

Min, K., Y. Son, C. Kim, Y. Lee, and K. Hong. 2007. Heat and moisture transfer from skin to environment through fabrics: A mathematical model. *Int. J. Heat Mass Tran.* 50: 5292–304.

Miraftab, M., R. Horrocks, and C. Woods. 1999. Carpet waste, an axpensive luxury we must do without! *AUTEX Res. J.* 1: 1–7.

Mizuno, O.K., K. Tsuzuki, Y. Ohshiro, and K. Mizuno. 2005. Effects of an electric blanket on sleep stages and body temperature in young men. *Ergonomics* 48: 749–57.

Möhring, U., D. Schwabe, and S. Hanus. 2011. Textiles for patient heat preservations during operations. In *Handbook of Medical Textiles*, Cambridge: Woodhead.

Morris, M.A. 1955. Thermal insulation of single and multiple layers of fabrics. *Text. Res. J.* 25: 766–73.

Morse, H.L., J.G. Thompson, K. Clark, K.A. Green, and C. Moyer. 1973. Analysis of the thermal response of protective fabrics, In *Technical Report, AFML-RT-73–17, Air Force Materials Information Service*, Virginia: Springfield.

Mukhopadhyay, A. and V.K. Midha. 2008. A review on designing the waterproof breathable fabrics. Part I: Fundamental principles and designing aspects of breathable fabrics. *J. Ind. Text.* 37: 225–62.

Murakami, S., S. Kato, and J. Zeng. 2000. Combined simulation of airflow, radiation and moisture transport for heat release from human body. *Build. Environ.* 35: 489–500.

Musat, R. and E. Helerea. 2009. Parameters and models of the vehicle thermal comfort. *Acta Univ. Sap. Electr. Mech. Eng.* 1: 215–26.

Muzet, A., J.P. Libert, and V. Candas. 1984. Ambient temperature and human sleep. *Cell. Mol. Life Sci.* 40: 425–29.

Nabovati, A., E.W. Llewellin, and A.C.M. Sousa. 2010. Through-thickness permeability prediction of three-dimensional multifilament woven fabrics. *Compos. Part A* 41: 453–63.

Natarajan, K. 2003. Air permeability of elastomeric fabrics as a function of uniaxial tensile strain, PhD thesis, North Carolina: NCSU.

Naydenov, K., D. Markov, T. Mustakov, A.K. Melikov, K. Arsen, T. Popov, P. Stankov, C.-G. Bornehag, G. Pichurov, Z.D. Bolashikov, and J. Sundell. 2006. Validation of self-reported health symptoms and housing characteristics in the ALLHOME project. In *Proceedings of the Healthy Buildings*, Lisbon, Portugal.

Naydenov, K., A. Melikov, D. Markov, P. Stankov, C.-G. Bornehag, and J. Sundell. 2008. A Comparison between occupants' and inspectors' reports on home dampness and their associations with the health of children: The ALLHOME study. *Build. Environ.* 43: 1840–49.

Nazarboland, M.A., X. Chen, J.W.S. Hearle, R. Lydon, and M. Moss. 2007. Effect of different particle shapes on the modelling of woven fabric filtration. *J. Inf. Comp. Sci.* 2: 111–18.

Nielsen, B. 1978. Physiology of thermoregulation during swimming. In *Swimming Medicine IV*, eds. B. Eriksson and B. Furberg, 297–303, Baltimore, MD: University Park Press.

Nocker, W. 2011. Evaluation of occupational clothing for surgeons. In *Handbook of Medical Textiles*, Cambridge: Woodhead.

Ogulata, R.T. 2006. Air permeability of woven fabrics. *J. Text. Apparel. Tech. Man* 5: 1–10.

Okubayashi, S., U.J. Griesser, and T. Bechtold. 2004. A kinetic study of moisture sorption and desorption on lyocell fibers. *Carbohyd. Polym.* 58: 293–99.

Paek, S.L. 1995. Effect of yarn type and twist factor on air permeability, absorbency and hand properties of open-end and ring-spun yarn fabrics. *J. Text. I.* 86: 581–9.

Pamuk, G. and F. Çeken. 2009. Research on the breaking and tearing strengths and elongation of automobile seat cover fabrics. *Text. Res. J.* 79: 47–58.

Park, S. and S. Jayaraman. 2003. Smart textiles: Wearable electronic systems. In *MRS Bulletin.* 28: 585–91.

Parsons, K. 2003. *Human Thermal Environments. The Effects of Hot, Moderate and Cold Environments on Human Health, Comfort and Performance*, London: Taylor & Francis.

Pause, B.H. 1995. Development of heat and cold insulating membrane structures with phase change material. *J. Coat. Fabrics* 25: 59–68.

Pause, B.H. 1998. Development of new cold protective clothing with phase change material. In *International Conference on Safety and Protective Fabrics*, Baltimore, MD, 78–84.

Peirce, F.T. and W.H. Rees. 1946. The transmission of heat trough textile fabrics. *J. Text. I* 37: 181–88.

Peirce, F.T. and F.T. Womersley. 1978. *Cloth Geometry*. Manchester, England: Textile Institute.

Phelan, F.R. and G. Wise. 1996. Analysis of transverse flow in aligned fibrous porous media. *Composites A* 27A: 25–34.

Pichurov, G. 2008. Gagge model, Tanabe model—Integration of CFD. In *Euro Academy on Ventilation and Indoor Climate* "CFD Based Design of Indoor Environment," 47–51. Sofia, Bulgaria: Avanguard Publishing.

Pichurov, G. 2009. Modelling and investigation of the aerodynamics of ventilated rooms and assesment of the thermal comfort, PhD thesis, Technical University, Sofia, Bulgaria (in Bulgarian).

Pichurov, G., R.A. Angelova, I. Simova, I. Rodrigo, and P. Stankov. 2014. CFD based study of thermal sensation of occupants using thermophysiological model. Part I: Mathematical model, implementation and simulation of the room air flow effect. *Int. J. Cloth Sci. Tech.* 26: 442–55.

Pichurov, G. and P. Stankov. 2013. Integration of thermophysiological body model in CFD. *Centr. Eur. J. Eng.* 3: 1–9.

Pierce, F. 1930. The handle of cloth as a measurable quantity. *Shirley Inst. Mem.* 9: 83–122.

Piekaar, H.W. and L.A. Clarenburg. 1967. Aerosol filters: Pore size distribution in fibrous filters. *Chem. Eng. Sci.* 22: 1399–407.

Plathner, P. and M. Woloszyn. 2002. Interzonal air and moisture transport in a test house. Experiment and modeling. *Build. Environ.* 37: 189–99.

Pluschke, P. (ed.). 2004. *The Handbook of Environmental Chemistry: Indoor Air Pollution*, Part F, Berlin, Germany: Springer-Verlag.

Randall, W.C. 1946. Quantitation and regional distribution of sweat glands. *J. Clin. Invest.* 25: 761–67.

Rief, S., E. Glatt, E. Laourine, D. Aibibu, C. Cherif, and A. Wiegmann. 2011. Modelling and CFD-simulation of woven textiles to determine permeability and retention properties, *AUTEX Res. J.* 11: 78–83.

Roberts, B.C., T.M. Waller, and M.P Caine. 2007. Thermoregulatory response to base-layer garments during treadmill exercise. *Int. J. Sports Sci. Eng.* 1: 29–38.

Roberts, J.W., W.S. Clifford, G. Glass, and P.G. Hummer. 1999. Reducing dust, lead, bacteria, and fungi in carpets by vacuuming. *Arch. Environ. Contam. Toxicol.* 36: 477–84.

Rode, C., R. Peuhkuri, K. Hansen, B. Tine, K. Svennberg, J. Arfvidsson, and T. Ojanen. 2002. Moisture buffer value of materials in buildings. In *Proceedings of the International 7th Nordic Symposium on Building Physics*, 108–15, Reykjavík, Sweden: Royal Institute of Technology.

Rushton, A. and P. Griffiths. 1971. Fluid flow in monofilament filter media. *Trans. Inst. Chem. Eng.* 49: 49–59.

Saeed, A., F.M. Anjum, R. Salim-Ur, A.M. Sheikh, and K. Farzana. 2008. Effect of fortification on physico-chemical and microbiological stability of whole wheat flour. *Food Chem.* 110: 113–19.

Salaun, F., E. Devaux, S. Bourbigot, and P. Rimeau. 2010. Development of phase change materials in clothing. Part I. *Text. Res. J.* 80: 195–205.

Saldaeva, E. 2010. *Through Thickness Air Permeability and Thermal Conductivity Analysis for Textile Materials*, PhD thesis, The University of Nottingham, Nottingham.

Satin, B., A.P. Gagge, and J.A.J. Stolwijk. 1970. Body temperatures and sweating during thermal transietnts caused by exercise, *J. Appl. Physiol.* 28: 318–27.

Saville, B.P. 1999. *Physical Testing of Textiles. Cambridge*: Woodhead Publishing.

Schacher, L. and A. Adolphe. 1997. Objective characterization of the thermal comfort of fabrics for car upholstery. In *Niches in the World of Textiles World Conference of the Textile Institute*, Manchester: Textile Institute, Vol. 2, 368–69.

Schacher, L., D.C. Adolphe, and J.-I. Drean. 2000. Comparison between thermal insulation and thermal properties of classical and microfibres polyester fabrics. *Int. J. Cloth. Sci. Tech.* 12: 84–95.

Schellen, L., M.G.L.C. Loomans, B.R.M. Kingma, M.H. de Wit, A.J.H. Frijns, and W.D. van Marken Lichtenbelt. 2013. The use of a thermophysiological model in the built environment to predict thermal sensation: coupling with the indoor environment and thermal sensation. *Build. Environ.* 59: 10–22.

Schiefer, H.F. 1943. Factors, related to thermal insulating factors of fabrics. *Text. Res. J.* 13: 21–6.

SDL Atlas. 2010. *M290 Moisture Management Tester, Instruction Manual*. Rock Hill, SC: SDL Atlas Inc.

Sefar. 2008. Precision woven synthetic filter fabrics. In *Sefar Filtration Solutions*, Customer Information 12. Buffalo, NY: Sefar Inc.

Seyam, A. and A. El-Shiekh. 1993. Mechanics of woven fabrics. *Text. Res. J.* 63: 371–78.

Shim, H. and E.A. McCullough. 2000. The effectiveness of phase change materials in outdoor clothing. In *Proceedings of the 1st European Conference on Protective Clothing*, Stockholm, Sweden.

Shishoo, R. 2000. Innovations in fibres and textiles for protective clothing. In *Proceedings of the 1st European Conference on Protective Clothing*, Stockholm, Sweden.

Sideroff, C.N. and T.Q. Dang. 2005. Validation of CFD for the flow around a computer simulated person in mixing ventilated room. In *Proceedings of the Conference on Indoor Air*, 1234–40. Beijing, China.

Simova, I., R.A. Angelova, and P. Stankov. 2009. Three dimensional simulation of the air-permeability of woven structures—Computational problems. In *Proceedings of International Conference of the Faculty of Power Engineering, EMF 2009*, 85–92, Sozopol, Bulgaria.

Singh, M.K. and A. Nigam. 2013. Effect of various ring yarns on fabric comfort. *J. Ind. Engineering*, Volume 2013, Article ID 206240, 7pp., http://dx.doi .org/10.1155/2013/206240.

Slater, K. 1985. *Human Comfort*, Springfield, IL: Charles C. Thomas Publisher.

Smith, B. and V. Bristow. 1994. Indoor air quality and textiles: An emerging issue. *Am Dyestuff Rep.* 83: 37–46.

Snycerski, M. and I. Frontczak-Wasiak. 2002. Influence of furniture covering textiles on moisture transport in a car seat upholstery package. *AUTEX Res. J.* 2: 126–31.

Sobera, M.P., C. Kleijn, P. Brasser, and E.A. Harry. 2004. A multi-scale numerical study of the flow, heat and mass transfer in protective clothing. In *Proceedings of Computational Science—ICCS 4th International Conference*, 637–44, Poland.

Song, G. 2011. Medical textiles and thermal comfort. In *Handbook of Medical Textiles*, ed. V.T. Bartels, Cambridge: Woodhead.

Song, G., R.L. Barker, H. Hamouda, A.V. Kuzenetsov, P. Chitrphiromsri, and R.V. Grimes. 2004. Modeling the thermal protective performance of heat resistant garments in flash fire exposures. *Text. Res. J.* 74: 1033–40.

Søresen, D.N. and L.K. Voigt. 2003. Modelling flow and heat transfer around a seated human body by computational fluid dynamics. *Build. Environ.* 38: 753–62.

Stankov, P. 1998. *Three-Dimensional Turbulent Flows in Heat and Mass Transfer Processes*, Doctor of Science thesis, Technical University of Sofia, Sofia, Bulgaria.

Stankov, P. 2003. Room air flow distribution—A key-factor for ventilation effectiveness. In *Proceedings of the 1st International Seminar "Healthy Buildings,"* 15–22, Sofia, Bulgaria: CERDECEN.

Stankov, P., M. Lozanova, D. Markov, and Y. Dinkov. 1994. Experimental investigation of rectangular turbulent jets. In *Proceedings of the International Symposium on Turbulence, Heat and Mass Transfer*, Lisbon, Portugal.

Stankov, P. and I. Simova. 2010. Numerical and experimental study on air jet systems applied in air supply devices. In *Proceedings of the International Conference "Mechanical Engineering in XXI Century,"* 47–51, Nis, Serbia.

Stolwijk, J.A.J. 1970. Mathematical model of thermoregulation. In *Physiological and Behavioral Temperature Regulation* 48: 703–21.

Stolwijk, J.A.J. and J.D. Hardy. 1966. Temperature regulation in man—A theoretical study. *Pfluegers Arch.* 291: 129–62.

Sun, G., H.S. Yoo, X.S. Zhang, and N. Pan. 2000. Radiant protective and transport properties of fabrics used by wildland firefighters. *Text. Res. J.* 70: 567–73.

Svennberg, K., L. Hedegaard, and C. Rode. 2004. Moisture buffer performance of a fully furnished room. In: *Whole Building Envelope IX*. Clearwater Beach, FL.

Svennberg, K. and L. Wadsö. 2008. Sorption isotherms for textile fabrics, foams and batting used in the indoor environment. *J. Text. I* 99: 125–32.

Takigami, H., G. Suzuki, Y. Hirai, Y. Ishikawa, M. Sunami, and S.-I. Sakai. 2009. Flame retardants in indoor dust and air of a hotel in Japan. *Environ. Int.* 35: 688–93.

Tan, Y.B., E.M. Crown, and L. Capjack. 1998. Design and evaluation of thermal protective flight suits. Part I: The design process and prototype development. *Cloth Text. Res. J.* 16: 47–55.

Tanabe, S.-I., K. Kobayashi, J. Nakano, Y. Ozeki, and M. Konishi. 2002. Evaluation of thermal comfort using combined multi-node thermoregulation (65MN) and radiation models and computational fluid dynamics (CFD). *Energ. Buildings* 34: 637–46.

Tarlochan, F. and S. Ramesh. 2005. Heat transfer model for predicting survival time in cold water immersion. *Biomed. Eng. Appl. Basis Commun.* 17: 159–66.

Taylor, M.A. 1991. *Technology of Textile Properties*, 3rd ed., London: Forbes Publications.

Thatcher, T.L. and D.W. Layton. 1995. Deposition, resuspension and penetration of particles within a residence. *Atmos. Environ.* 29: 1487–97.

Tokarska, M. 2004. Neural model of the permeability features of woven fabrics. *Text. Res. J.* 74: 1045–48.

Torvi, D.A. 1997. *Heat Transfer in Thin Fibrous Materials under High Heat Flux Conditions*, PhD Thesis, Canada: University of Alberta.

Travers, E.B. and N.F. Olsen. 1982. Effect of air permeability on smoldering characteristics of cotton upholstery fabrics. *Text. Res. J.* 52: 598–604.

Tudor-Locke, C., B.E. Ainsworth, T.L. Washington, and R. Troiano. 2011. Assigning metabolic equivalent values to the 2002 census occupational classification system. *J. Phys. Act. Health* 8: 581–86.

Ukponmwan, J.O. 1993. The thermal-insulation properties of fabrics. *Text. Prog.* 24: 1–54.

Ülkü, S., D. Balköse, T. Çağa, F. Özkan, and S. Ulutan. 1998. A study of adsorption of water vapour on wool under static and dynamic conditions. *Adsorption* 4: 63–73.

Umbach, K. 1986. Comparative thermophysiological tests on blankets made from wool and acrylic-fibre-cotton blends. *J. Text. I* 77: 212–22.

Umbach, K. 1993. Moisture transport and wear comfort in microfibers fabrics. *Melliand Textillberichte* 74: 173–78.

U.S. Environmental Protection Agency. 1989. Report to Congress on indoor air quality: Volume 2. EPA/400/1-89/001C. Washington, DC.

Van Langenhove, L. and C. Hertleer. 2004a. Smart clothing: A new life. *Int. J. Cloth. Sci. Tech.* 16: 63–72.

Van Langenhove, L. and C. Hertleer. 2004b. Smart textiles in vehicles: A foresight. *JTATM* 3: 1–6.

Van Treeck, C., J. Frisch, M. Pfaffinger, and E. Rank. 2009. Thermal comfort analysis using a parametric manikin model for interactive simulation. *J. Build Perform. Sim.* 2: 233–50.

Verleye, B. 2008. *Computation of the Permeability of Multi-Scale Porous Media with Application to Technical Textiles,* PhD thesis, Catholic University of Leuven, Belgium.

Verleye, B., M. Klitz, R. Groce, D. Roose, S.V. Lomov, and I. Verpoest. 2007. Computation of the permeability of textiles with experimental validation for monofilament and non crimp fabrics. *Stud. Comp. Intell.* 55: 93–110.

Verleye, B., G. Morren, S.V. Lomov, H. Sol, I. Verpoest, and D. Roose. 2009. User-friendly permeability predicting software for technical textiles. *RJTA* 13: 19–27.

Vlaović, D. and G. Župčić. 2012. Thermal comfort while sitting on office chairs— Subjective evaluations. *Drvna Industrija* 63: 263–70.

Wadsö, L., K. Svennberg, and A. Dueck. 2004. An experimentally simple method for measuring sorption isotherms. *Drying Tech.* 22: 2427–40.

Wagner, R.D., R.C. Leslie, and F. Schlaeppi. 1985. Test methods for determining colourfastness to light. *Text. Chem. Color.* 17: 17–27.

Walz, M. 2011. Occupational clothing for nurses. In *Handbook of Medical Textiles,* ed. V.T. Bartels, Cambridge: Woodhead.

Wang, J., J.K. Carson, M.F. North, and D.J. Cleland. 2008. A new structural model of effective thermal conductivity for heterogeneous materials with co-continuous phases. *Int. J. Heat Mass. Tran.* 51: 2389–97.

Wang, Q., B. Maze, and H.V. Tafreshi. 2007. On the pressure drop modeling of monofilament-woven fabrics. *Chem. Eng. Sci.* 62: 4817–21.

Wang, X. and Z. Liu. 2004. The Forchheimer equation in two-dimensional percolation porous media. *Physica A.* 337: 384–88.

Wang, Y., J. Moss, and R. Thisted. 1992. Predictors of body surface area. *J. Clin. Anesth.* 4(1): 4–10.

Warfield, C.L. 1987. Upholstered furniture: Results of a consumer wear study. *Text. Res. J.* 57: 192–99.

Watt, I.C. and R.L. Darcy. 1979. Water-vapour adsorption isotherms of wool. *J. Text. I.* 70: 298–307.

Wendt, J.F. 1992. *Computational Fluid Dynamics CFD,* Germany: Springer-Verlag.

Wersteeg, H.K. and W. Malalsekera. 1995. *An Introduction to CFD,* England: Longman.

Whitaker, S.A. 1977. A theory of drying in porous media. *Adv. Heat. Trans.* 13: 119–203.

Whitmore, R.W., F.W. Immerman, D.E. Camann, A.E. Bond, R.G. Lewis, and J.L. Schaum. 1994. Non-occupational exposures to pesticides for residents of two US cities. *Arch. Environ. Contain. Toxicol.* 26: 47–59.

Wilbik-Halgas, B., R. Danych, B. Wiecek, and K. Kowalski. 2006. Air and water vapour permeability in double-layered knitted fabrics with different raw materials. *Fibers Text. East. Eur.* 14: 77–80.

Wilson, C.A. and R.M. Laing. 1995. Investigation of selected tactile and thermal characteristics of upholstery fabrics. *Cloth. Text. Res. J.* 13: 200–7.

Williams, S.D. and D.M. Curry. 1992. *Thermal Protection Materials: Thermophysical Property Data*, NASA Reference Publication, NASA, US.

Winslow, C.E.A., L.P. Herrington, and A.P. Gagge. 1936. A new method of partitional calorimetry. *Am. J. Appl. Physiol.* 116: 641–55.

Winslow, J.D., L.P. Herrington, and A.P. Gagge. 1937. Physiological reactions of the human body to varying environmental temperatures. *Am. J. Appl. Physiol.* 120: 1–22.

Wissler, E.H. 1961. Steady-state temperature distribution in man. *J. Appl. Physiol.* 16: 734–40.

Wissler, E.H. 1963. An analysis of factors affecting temperature levels in the nude human. In *Temperature Its Measurements and Control in Science and Industry*, New York: Reinholds Publishing.

Wong, C.C. 2006. *Modelling the effect of textile preform architecture on permeability*, PhD thesis, University of Nottingham, UK.

Woods, J.E., D.T. Braymen, R.W. Rasmussen, G.L. Reynolds, and G.M. Montag. 1986. Ventilation requirements in hospital operating rooms—Part I: Control of airborne particles, *ASHRAE Trans.* 92(2A): 396–426.

Worfolk, J. 1997. Keep elders warm. *Geriatr. Nurs.* 18: 7–11.

Wyon, D. 1994. Current indoor climate problems and their possible solution. *Indoor Built Environ.* 3: 123–9.

Xiao, X. 2012. *Modelling the Structure-Permeability Relationship for Woven Fabrics*, PhD thesis, The University of Nottingham, Nottingham.

Xu, G. and F. Wang. 2005. Prediction of the permeability of woven fabrics. *J. Ind. Text.* 34: 243–54.

Yoon, H.N. and A. Buckley. 1984. Improved Comfort polyester. Part I: Transport properties and thermal comfort of polyester/cotton blend fabrics. *Text. Res. J.* 54: 289–98.

Zhou, X., Z. Lian, and L. Lan. 2013. An individualized human thermoregulation model for Chinese adults. *Build. Environ.* 70: 257–65.

Zimmerli, T. 2000. Past, present and future trends in protective clothing. In *Proceedings of the 1st European Conference on Protective Clothing*, Stockholm, Sweden.

Zupin, Z., A. Hladnik, and K. Dimitrovski. 2012. Prediction of one-layer woven fabrics air permeability using porosity parameters. *Text. Res. J.* 82: 117–28.

Index